Statistics is Easy!

Statistics is Easy!
Dennis Shasha and Manda Wilson
www.morganclaypool.com

ISBN: 9781598297775   paperback

ISBN: 9781598297782   ebook

DOI: 10.2200/S00142ED1V01Y200807MAS001

A Publication in the Morgan & Claypool Publishers' series

SYNTHESIS LECTURES ON MATHEMATICS AND STATISTICS #1

Lecture #1

Series Editor: Steven G. Krantz, Washington University St. Louis

10 9 8 7 6 5 4 3 2 1

Statistics is Easy!

Dennis Shasha
Department of Computer Science
Courant Institute of Mathematical Sciences
New York University

Manda Wilson
Bioinformatics Core
Computational Biology Center
Memorial Sloan-Kettering Cancer Center

SYNTHESIS LECTURES IN MATHEMATICS AND STATISTICS #1

MORGAN & CLAYPOOL PUBLISHERS

ABSTRACT

Statistics is the activity of inferring results about a population given a sample. Historically, statistics books assume an underlying distribution to the data (typically, the normal distribution) and derive results under that assumption. Unfortunately, in real life, one cannot normally be sure of the underlying distribution. For that reason, this book presents a distribution-independent approach to statistics based on a simple computational counting idea called resampling.

This book explains the basic concepts of resampling, then systematically presents the standard statistical measures along with programs (in the language Python) to calculate them using resampling, and finally illustrates the use of the measures and programs in a case study. The text uses junior high school algebra and many examples to explain the concepts. The ideal reader has mastered at least elementary mathematics, likes to think procedurally, and is comfortable with computers.

All code was tested using Python 2.3.
www.python.com

Note: When clicking on the link to download the individual code or input file you seek, you will in fact be downloading all of the code and input files found in the book. From that point on, you can choose which one you seek and proceed from there. In order to download the data file NMSTATEDATA4.2[1] located in Chapter 5, go to: http://www.morganclaypool.com/userimages/ContentEditor/1214542013584/NMStateData.zip.

ACKNOWLEDGEMENTS

All graphs were generated using GraphPad Prism version 4.00 for Macintosh, GraphPad Software, San Diego California USA, www.graphpad.com. Dennis Shasha's work has been partly supported by the U.S. National Science Foundation under grants IIS-0414763, DBI-0445666, N2010 IOB-0519985, N2010 DBI-0519984, DBI-0421604, and MCB-0209754. This support is greatly appreciated. The authors would like to thank Radha Iyengar, Rowan Lindley, and Jonathan Jay Monten for their helpful reading. In addition, copyeditor Sara Kreisman and compositor Tim Donar worked under tight time pressure to get the book finally ready.

Introduction

Few people remember statistics with much love. To some, probability was fun because it felt combinatorial and logical (with potentially profitable applications to gambling), but statistics was a bunch of complicated formulas with counter-intuitive assumptions. As a result, if a practicing natural or social scientist must conduct an experiment, he or she can't derive anything from first principles but instead must pull out some dusty statistics book and apply some formula or use some software, hoping that the distribution assumptions allowing the use of that formula apply. To mimic a familiar phrase: "There are hacks, damn hacks, and there are statistics."

Surprisingly, a strong minority current of modern statistical theory offers the possibility of avoiding both the magic and assumptions of classical statistical theory through randomization techniques known collectively as *resampling*. These techniques take a given sample and either create new samples by randomly selecting values from the given sample with replacement, or by randomly shuffling labels on the data. The questions answered are familiar: How accurate is the measurement likely to be (confidence interval)? And, could it have happened by mistake (significance)?

A mathematical explanation of this approach can be found in the well written but still technically advanced book *Bootstrap Methods and their Application* by A. C. Davison and D. V. Hinkley. We have also found David Howell's web page (http://www.uvm.edu/~dhowell/StatPages/Resampling/Resampling.html) extremely useful. We will not, however, delve into the theoretical justification (which frankly isn't well developed), although we do note that even formula-based statistics is theoretically justified only when strong assumptions are made about underlying distributions. There are, however, some cases when resampling doesn't work. We discuss these later in Chapter 2.

Note to the reader: We attempt to present these ideas constructively, sometimes as thought experiments that can be implemented on a computer. If you don't understand a construction, please reread it. If you still don't understand, then please ask us. If we've done something wrong, please tell us. If we agree, we'll change it and give you attribution.

Contents

The Basic Idea

Suppose you want to know whether a coin is fair[1]. You toss it 17 times and it comes up heads all but 2 times. How might you determine whether it is reasonable to believe the coin is fair? (A fair coin should come up heads with probability 1/2 and tails with probability 1/2.) You could ask to compute the percentage of times that you would get this result if the fairness assumption were true. Probability theory would suggest using the binomial distribution. But, you may have forgotten the formula or the derivation. So, you might look it up or at least remember the name so you could get software to do it. The net effect is that you wouldn't understand much, unless you were up on your probability theory.

The alternative is to do an experiment 10,000 times, where the experiment consists of tossing a coin that is known to be fair 17 times and ask what percentage of times you get heads 15 times or more (see Fig. 1.1). When we ran this program, the percentage was consistently well under 5 (that is, under 5%, a result often used to denote "unlikely"), so it's unlikely the coin is in fact fair. Your hand might ache from doing this, but your PC can do this in under a second.

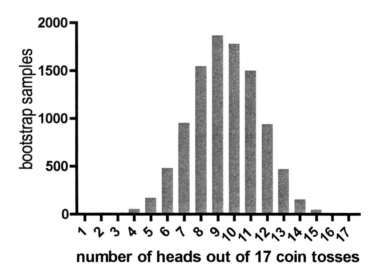

Figure 1.1:   Coin toss.

---

1.   We take this coin example from David Howell.
     See http://www.uvm.edu/~dhowell/StatPages/Resampling/philosophy.html.

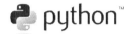 python™

download code and input files here:
http://www.morganclaypool.com/userimages/ContentEditor/1216676640846/Statistics_is_Easy_Codes.zip

Here is an example run of the Coinsig.py code:

```
9 out of 10,000 times we got at least 15 heads in 17 tosses.
Probability that chance alone gave us at least 15 heads in 17 tosses is 0.0009 .
```

Here is a second example.

Imagine we have given some people a placebo and others a drug. The measured improvement (the more positive the better) is:

```
Placebo: 54 51 58 44 55 52 42 47 58 46
Drug: 54 73 53 70 73 68 52 65 65
```

As you can see, the drug seems more effective on the average (the average measured improvement is nearly 63.7 (63 2/3 to be precise) for the drug and 50.7 for the placebo). But, is this difference in the average real? Formula-based statistics would use a *t*-test which entails certain assumptions about normality and variance; however, we are going to look at just the samples themselves and *shuffle* the labels.

The meaning of this can be illustrated in the following table—in which we put all the people— labeling one column 'Value' and the other 'Label' (P stands for placebo, D for drug).

Value	Label
54	P
51	P
58	P
44	P
55	P
52	P
42	P
47	P
58	P
46	P
54	D

73	D
53	D
70	D
73	D
68	D
52	D
65	D
65	D

Shuffling the labels means that we will take the P's and D's and randomly distribute them among the patients. (Technically, we do a uniform random permutation of the label column.)

This might give:

Value	Label
54	P
51	P
58	D
44	P
55	P
52	D
42	D
47	D
58	D
46	D
54	P
73	P
53	P
70	D
73	P
68	P
52	D
65	P
65	D

In Fig. 1.2, we can then look at the difference in the average P value vs. the average D value. We get an average of 59.0 for P and 54.4 for D. We repeat this shuffle-then-measure procedure 10,000 times and ask what fraction of time we get a difference between drug and placebo greater than or equal to the measured difference of 63.7 - 50.7 = 13. The answer in this case is under 0.001. That is less than 0.1%. Therefore, we conclude that the difference between the averages of the samples is real. This is what statisticians call *significant*.

Let's step back for a moment. What is the justification for shuffling the labels? The idea is simply this: if the drug had no real effect, then the placebo would often give more improvement than the drug. By shuffling the labels, we are simulating the situation in which some placebo measurements replace some drug measurements. If the observed average difference of 13 would be matched or even exceeded in many of these shufflings, then the drug might have no effect beyond the placebo. That is, the observed difference could have occurred by chance.

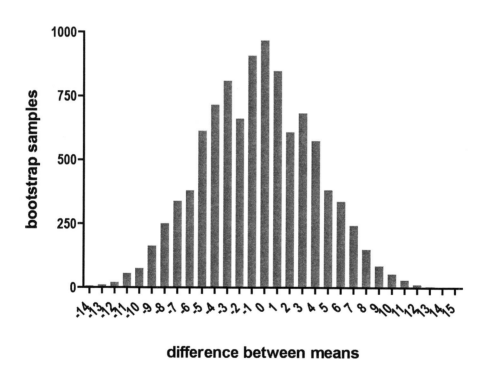

Figure 1.2: Difference between means.

To see that a similar average numerical advantage might lead to a different conclusion, consider a fictitious variant of this example. Here we take a much greater variety of placebo values: 56 348 162 420 440 250 389 476 288 456 and simply add 13 more to get the drug values: 69 361 175 433 453 263 402 489 301 469. So the difference in the averages is 13, as it was in our original example. In tabular form we get the following.

Value	Label
56	P
348	P
162	P
420	P
440	P
250	P
389	P
476	P
288	P
456	P
69	D
361	D
175	D
433	D
453	D
263	D
402	D
489	D
301	D
469	D

This time, when we perform the 10,000 shufflings, in approximately 40% of the shufflings; the difference between the D values and P values is greater than or equal to 13. So, we would conclude that the drug may have no benefit — the difference of 13 could easily have happened by chance.

All code was tested using Python 2.3.
www.python.com

download code and input files here:
http://www.morganclaypool.com/userimages/ContentEditor/1216676640846/Statistics_is_Easy_Codes.zip

Here is an example run of the Diff2MeanSig.py code, using the first data set from this example (found on p. 2) as input:

```
Observed difference of two means: 12.97
7 out of 10,000 experiments had a difference of two means greater than or
equal to 12.97 .
The chance of getting a difference of two means greater than or equal to 12.97 is
0.0007.
```

In both the coin and drug case so far, we've discussed **statistical significance**. Could the observed difference have happened by chance? However, this is not the same as importance, at least not always. For example, if the drug raised the effect on the average by 0.03, we might not find this important, even if the result is statistically significant. That is, the first question you should ask when someone tells you an effect is statistically significant is: "Yes, but how large is the effect?" Perhaps what is being measured here is survival. Say average survival on the placebo is 5 years, and that the drug increases survival on average by 3 days. The difference between 5 years and 5 years and 3 days may be significant, but it is not a large effect.

To get a feeling for this question of importance, we will use the notion of a *confidence interval*. Intuitively, the confidence interval of an imperfectly repeatable measurement is defined by the range of values the measurement is likely to take. In resampling statistics as in traditional statistics, this range is commonly defined as the middle 90% (or sometimes 95%) of the possible values. If you've been following carefully so far, you will guess that the set of possible values will be based on repeated random samples of some sort. In the drug case, we will take many samples from the patient data we have and then look at the difference between the average drug improvement and the average placebo improvement. We'll look at the range of these differences and compute the confidence interval. This technique is called *bootstrapping*.

Here's the method: we create new samples of the same size as the original by choosing values from the original sample "uniformly at random and with replacement".

Let's break down the phrase. "Uniformly at random" means each new sample element is chosen from the original sample in such a way that every original sample element has the same chance of being picked. "With replacement" means that even though an original sample element has been picked, its chance of getting picked again remains the same. Simply put, in forming a new sample (called a bootstrap sample), we choose uniformly at random on the original sample and may choose some elements twice or more and some elements no times at all.

Let's recall our original data regarding drugs and placebos:

```
Placebo: 54 51 58 44 55 52 42 47 58 46
The average is: 50.7.


Drug: 54 73 53 70 73 68 52 65 65
The average is: 63.7.
```

> **NOTE:**
> As we will see later in Ch. 3, this is in fact not enough data to justify the confidence interval procedure, but is used for easier illustration.

We subtract the placebo average from the drug average, yielding 63.7 - 50.7 = 13.

Our question now will be: "What is the 90% confidence interval of difference in the averages between the drug patients and placebo?" We answer this with experiments of the form: take a bootstrap sample of the placebo patients and compute the average; take a bootstrap sample of the drug patients and compute the average; then subtract the placebo average from the drug average. When we do this 10,000 times (the rule of thumb for bootstrapping is 1,000 times, but to increase the probability of capturing a wider range of values, we advocate increasing this to 10,000), we get many differences.

Here is a typical experiment in which a bootstrap sample of the placebo values is (note that 54 and 55 are repeated a few times, but 52 never appears):

```
55 54 51 47 55 47 54 46 54 54
The average is: 51.7.
```

Here is a bootstrap of the drug values:

```
68 70 65 70 68 68 54 52 53
The average is: 63.1.
```

We subtract the placebo average from the drug average, yielding 63.1 - 51.7 = 11.4.

When we repeated such an experiment 10,000 times and performed the subtraction each time, the lowest difference was -0.46 (the placebo is a tiny bit more effective than the drug). The highest was 23.4 (the drug is much more effective than the placebo). A more interesting range is the value 5% from the lowest and 95% from the lowest (percentile 5% and 95%). That is, arrange the differences in sorted order from smallest to largest and pick the differences that are at position 500 (500 is 5% of 10,000) and the difference at position 9,500. That is the 90% confidence interval. In our experiments, this yields a range of 7.5 to 18.1. That is, 90% of the time, drugs should yield a value that is 7.5 to 18.1 more than the placebo.

All code was tested using Python 2.3.
www.python.com

download code and input files here:
http://www.morganclaypool.com/userimages/ContentEditor/1216676640846/Statistics_is_Easy_Codes.zip

Here is an example run of the Diff2MeanConf.py code:

```
Observed difference between the means: 12.97
We have 90.0 % confidence that the true difference between the means is between:
7.81 and 18.11
```

Confidence intervals may vary vastly more for social/cultural phenomena than for physical/ biological ones. Suppose we have 20 people and we compute their average income. Most have an annual income in the multi-thousand dollar range, but one person has an income of a billion dollars. The average will therefore be something like $50 million, even though most people don't make that much.

What we might be interested in is how far that average value varies if this were in fact a typical sample. Using **bootstrapping** we might have incomes (measured in thousands) as follows:

```
200 69 141 45 154 169 142 198 178 197 1000000 166 188 178 129 87 151 101 187 154
```

This gives an average of 50,142 thousands.

Now if we use the bootstrap, here is another sample:

```
151 154 166 188 154 101 1000000 129 188 142 188 129 142 188 151 87 200 178 129 166
```

This has an average of 50,146 thousands.

Another one:

```
154 87 178 151 178 87 154 169 187 129 166 154 154 166 198 154 141 188 87 69
```

This has an average of only 147 thousands (because the billionaire is missing).

Another one:

```
69 166 169 142 188 198 154 45 187 166 87 154 1000000 87 151 166 101 154 1000000 166
```

This has an average of 100,127 thousands because the billionaire is present twice.

The net effect is that we are going to get a wide variety of averages. In fact, when we ran the bootstrap 10,000 times on our PC, we obtained a low average of 114 thousands and a high average of 300,118 thousands. But these highs and lows are not so interesting because they vary a lot depending on how many times the billionaire happens to appear. A more interesting range is the value 5% from the lowest and 95% from the lowest (percentile 5% and 95%). That is, arrange the averages in sorted order from smallest to largest and pick the average that is at position 500 (500 is 5% of 10,000) and the average at position 9,500.

On our PC, the 5th percentile is 138 thousands and the 95th percentile is 200,130 thousands. Because 95 - 5 = 90, this is the 90% confidence interval. Because of this vast range, we'd probably conclude that the average is not very informative.

When data has such extreme *outliers*, there are sometimes reasons for ignoring them, such as a faulty meter reading. If there is a good reason to ignore the billionaire in this case, then we get a 90% **confidence interval** of about 132 to 173 thousands, a much more narrow range of expected values. Unfortunately, outliers are ignored incorrectly sometimes, so one must be careful. Also, as Nassim Taleb points out persuasively, outliers are much more common in human constructs (like income or inflation) than in natural phenomena (like rainflow; see http://

www.edge.org/3rd_culture/taleb04/taleb_index.html). He gives a particularly convincing example: in the early 1920s, inflation in Germany caused the exchange rate from German Marks to U.S. dollars to go from 3 to a dollar to 4 trillion to a dollar (that is a statistical impossibility under the normal distribution assumption, which illustrates why blind application of the normal distribution can be dangerous).[2]

If you've been following this carefully, you might now wonder "If I have a **confidence interval**, what more does **significance** bring to the party?" To answer this intuitively, consider a simple example in which you have just one element of group A having value 50 and one element of group B having value 40. The confidence interval using replacement will say that the difference is always 10. But intuitively this is way too little data. The significance test (in which one permutes the group labels) will show that half the time one will get, just by random chance, a difference as big as the observed one.

If all this seems easy, that's good. Several studies indicate that resampling is easier for students to learn and they get correct answers more often than using classical methods (see http://www.resample.com/content/teaching/texts/i-1reslt.txt).

---

2. Why do we prefer **confidence intervals** based on the bootstrapping method to traditional confidence intervals based on the standard deviation of the data? First, because we don t want to have to make the assumption that the underlying distribution is normal. Second, because many distributions are in fact skewed. For example, if we want to know the average salary of a population and we know the salaries of a sample, we expect salaries to be positive, whereas the average less the standard deviation might in fact be negative. Bootstrapping looks at the data that is present.

Pragmatic Considerations When Using Resampling

This material is based on Sec. 2.6 of the Davison and Hinkley book *Bootstrap Methods and their Application*, and Philip Good in his book *Resampling Methods*. It concludes with a short primer on statistical jargon to which you may want to refer in your reading.

1. **Bootstrapping** may underestimate the size of the **confidence interval** when the sample is small (as a rule of thumb, according to Philip Good, under 100). In that case, a **significance** test may work better. Significance/**shuffle**/permutation tests (when one shuffles the label) can be used with as few as three data points of each type. This implies that there may be cases when one wants to reformulate a question as a significance test when there are few data items. For example, to evaluate whether a treatment T1 encourages, say, growth more than treatment T2, one could use three strategies:

    i. Find the 95% **bootstrap confidence interval** of the difference in the means between T1 and T2.

    ii. Label each data point with the treatment it came from. Then do a **significance** test by shuffling the labels and seeing how often the difference in the means exceeds the observed difference.

    iii. **Shuffle** the labels as in (ii) but choose the data points by sampling with replacement. (A hybrid significance approach.)

    Option (i) implicitly assumes there is a difference and simply attempts to find the confidence interval of its magnitude. As such, it does what statisticians call "starting with the **alternative hypothesis**." Options (ii) and (iii) start with the **null hypothesis** (the treatment doesn't matter). The *p*-**value** calculated in their case is the probability that the treatment doesn't matter but that the difference one sees is as great as that observed. That's why they constitute a significance approach.

    To see why (i) differs in its implicit assumption from the other two, suppose that we had the degenerate case of just one data value from each treatment, e.g., the T1 value is 55 and the T2 value is 48. Option (i) would always find the same difference between the means, in this case 7, so the 95% confidence interval would range from 7 to 7. Options (ii) and (iii) would reshuffle the labels so about half the time the labels would be reversed, so the *p*-**value** of a difference this great would be about 0.5, obviously not significant.

    Pragmatically, therefore, you should start by deciding whether the null hypothesis should be rejected before measuring the magnitude of the difference. If the variances of the two samples are different, then using (iii) is probably best.

2. **Bootstrapping** should not be used to find the maximum value of an underlying population (e.g., to try to find the tallest person in Holland, it would not do to sample 1,000 Dutch people). This holds also if one wants to find the kth largest value. On the other hand, bootstrapping can be used to estimate the cutoff demarking the 10% largest values (also known as the highest decile) and to get **confidence intervals** for those values.

3. **Bootstrapping** also should be used with great caution whenever there are **outliers** in a sample, e.g., values very different from the others that can radically change the evaluation of the statistic. If outliers are a problem, then a *rank transformation* might work better (e.g., convert every salary to a salary rank as in 1 for the highest, 2 for the second highest, etc.).

To locate the outliers (see http://en.wikipedia.org/wiki/Outlier), start by computing the difference between the third quartile value q_3 (the value below which 75% of the data lies) and the first quartile value q_1 (the value below which 25% of the data lies). Call this the inter-quartile range IQR. An outlier is any value less than $q_1 - 3IQR$ or greater than $q_3 + 3IQR$. A temptation is to discard outliers. As mentioned earlier, throwing out outliers blindly can lead one to miss important phenomena (holes in the ozone layer, people with extraordinarily high income, stock market crashes, and so on).

More sophisticated statistical approaches for handling outliers use a technique known as "Robust Statistics" described nicely in the texts by Huber or Hampel et al. The consensus among the professionals, however, is that outliers should be thought about carefully or transformed into ranks rather than statistically eliminated.

4. Neither **bootstrapping** nor **significance** should be used when the sample is not a representative one. For example, if you are doing a survey and the yield (the fraction of people called who respond) is very low, then the sample may be biased. Whether it is or not depends on whether the reason for refusal is relevant to the quantity you're measuring. For example, if people opt out of a medical study because they find out they are getting a placebo, then any conclusions about outcomes may be biased in favor of the drug. It is the possible lack of representativeness of exit polling that may explain why voting results at the ballot box seem to differ from exit polling estimates. Lack of representativeness is a problem for any statistical technique.

Sometimes there can be experimental bias. Philip Good observes that when placing subjects into treatment pools, researchers tend to put the most promising ones into the first pool (e.g., the healthiest plants into the first garden). It is essential, therefore, to randomly distribute values among the various treatment pools. The pseudocode if you have n subjects and k treatments is:

```
for subject 1 to n
  choose a number x and at random between 1 and k
  put the subject in treatment x unless x is full
  in which case you choose again until you find a treatment
     that is not full
end for
```

5. **Resampling** should be used with care when the data exhibits serial dependence. For example, permuting individual elements in a musical melody or stock price time series may give erroneous results for the simple reason that the value at each time point depends to a great extent on the value at the previous time point. In such a case, the basic assumption of resampling—that values in the sample don't depend on the values around them—doesn't hold. A practical fix for this, if the sample is large enough, is to take non-overlapping long consecutive subsequences of data and then do resampling on entire subsequences. This technique preserves the local

dependencies of the data except at the borders of the subsequences. Tomislav Pavlicic of DE Shaw offered us the following case history: "I successfully used resampling for mutually dependent data. It did require creating 50,000 virtual subjects each with 100 mutually dependent data points and then performing resampling on non-overlapping (longish) subsamples. It worked! (The alternative would have required the kind of expertise in combinatorics which was way above my head.)"

Victor Zhu pointed out another (advanced) technique for evaluating significance when there is serial dependence: "One technique I have seen in permutation for time series is to calculate the wavelet coefficients of each time series, to permute the labels associated with each coefficient [so the first wavelet coefficient from time series x may become the first wavelet coefficient from time series y] and then to reconstruct the several time series. This is a way to test the **significance** of correlations.

"Another question I have often encountered in interviews is how to tell if one trading strategy is better than another, or is it just due to luck. Try a **shuffle** test."

6. *Rank transformations* can be very useful sometimes. For example, suppose you are calculating the economic value of going to college. You look at the wealth of all people who graduated from college as compared to all people who didn't. The trouble is that a few **outliers** such as Bill Gates might **significantly** influence the result. So, you might decide to order all wealths and replace each person's wealth by his or her rank (say that Gates would be 1). Then the question is: what is the average rank of college graduates vs. non-college graduates? If lower for college graduates, then college has some economic benefit in the sense that college graduates tend to be among the higher earners. (Remember that rank 1 is the richest individual.) In detail, this analysis would involve associating with each individual a rank and a label (college or non-college). Compute the average rank of the individuals having each label. Then perform a significance test that involves randomly *shuffling* the labels and measuring the difference in the means. By doing this repeatedly, you will determine how often the average college graduate rank is less than the average non-college rank. This would give a p-value (or significance value). Note that this transformation addresses the question: "Do college graduates tend to be among the higher earners?" rather than "Do college graduates have a higher mean salary?"

Transformations can also involve simple algebra. For example, if you want to test the proposition that Dutch men are more than five centimeters taller than Italian men, you can make the **null hypothesis** be that Dutch men are five centimeters taller. To do this, first subtract five centimeters from the heights of the Dutch men in your sample. Now you have men represented by a pair (height, nationality), where the heights have been adjusted for the Dutch men. The null hypothesis is now that the average height difference should not differ from 0. Suppose that the average (adjusted) height among the Dutch men is in fact greater than the average height among the Italian men. Suppose that after the transformation Dutch men were on average still 4 centimeters taller (so 9 centimeters on the original). Suppose further that under the null hypothesis after the transformation, shuffling the nationality labels seldom yields this outcome. If that is true, then the observed extra 4 centimeters has a low p-**value** so the null hypothesis is unlikely to hold. That is, the height difference is likely to be greater than 5 centimeters.

7. For the most part, we concentrate on the use of **resampling** to estimate the **significance** of a result (using **shuffling**) or the size of a result (using **bootstrapping**) regardless of the under-

lying distribution. An important alternative use of resampling is to estimate the probability that a sample comes from a particular distribution. We did this in Fig. 1.1 when we asked how likely a set of 17 flips of a fair coin would be to yield 15 heads or more. The basic concept is to use bootstrapping from the assumed distribution and then to see how often the observed outcome materializes.

8. The *power of a test* is the probability that the test will reject the **null hypothesis** if the null hypothesis is false. The power increases the larger the size of the sample. Given a desired power and either a distribution D reflecting your prior belief or an *empirical distribution* (a distribution based on an experiment), you can determine the size of the sample you need to achieve that power. The idea is simple: assume that D is true and say you are testing some statistic such as the mean M and you want to know how big a sample you'll need to conclude 90% of the time (i.e., with a power of 90%) that $M > 50$ with a ***p*-value** < 0.05.

Specify a sample size S. We will see whether S is large enough to give us the power we want to achieve the **alternate** (i.e., non-null) **hypothesis**, that $M > 50$.

```
Try the following 1,000 times:
    draw 10,000 samples of size S from the empirical distribution
                with replacement
        see whether M is over 50 at least 95% of the time
        If so, count this as a reject of the null hypothesis
```

The power is the fraction rejects/1,000. If the power is too low (less than 0.9 in this example), you'll need a bigger sample.

For example, suppose your empirical distribution indicates that the data has a probability of 0.4 of having the value 40 and a probability of 0.6 of having the value 70. You want a power of 0.8 that you will find a mean greater than 50 with a *p*-value under 0.05.

```
Try 1,000 times with a sample size S:
    take 10,000 samples of size S from this distribution
    if 95% or more of the time, one gets a mean over 50,
        then the number of rejects increases by 1
```

The sample size is big enough if the number of rejects divided by $1,000 > 0.8$. Otherwise try a larger S and repeat. See also http://www.pubmedcentral.nih.gov/articleren?der.fcgi?artid=421748.

Note that we have selected a single sample mean as our test statistic, but this method can be used for multiple data sets and other statistics. The point is to use this same strategy: assume a distribution then try a sample size by repeatedly drawing a sample from that distribution of that size.

For example, suppose you believe that treatment T1 obeys some distribution D1 and T2 obeys D2 and the mean of D1 is greater than the mean of D2. You want to know whether taking a sample of S points from each will often cause you to reject the null hypothesis that the mean of T1 equals the mean of T2 with a *p*-value of 0.05.

```
Do the following 1,000 times
   Do 10,000 times: take a sample of size S from presumed
   distribution D1 and another sample of size S from presumed distribution D2
   and see whether mean of the sample from D1 >  mean of the sample from D2
   If this is true 95% of the time, then count this as a reject of the null
hypothesis.
```

The power is the number of rejects divided by 1,000.

In order to find the correct sample size more efficiently, we may perform the outer loop far fewer times. If after 100 iterations of the outer loop the power looks way too low, then we increase the sample size immediately.

9. Multi-factor designs and *blocking*[1]. You want to test different dosages of a fertilizer on a field. You are worried, however, that the properties of the field might swamp the effect of the fertilizer. On the other hand, it is not practical to give two plants that sit right next to one another a different fertilizer, so large scale randomization is not possible. For this reason, you might use a *Latin square design* so similar dosages (Lo, Med, Hi) of fertilizer are not applied to neighboring fields.

	east	**—**	**west**
north	Hi	Med	Lo
—	Lo	Hi	Med
south	Med	Lo	Hi

This is better than a randomized design because a randomized design on a three-by-three square might give a very biased outcome like:

	east	**—**	**west**
north	Hi	Hi	Lo
—	Lo	Hi	Med
south	Med	Lo	Med

When doing a **shuffle** test in this setting, it is helpful to shuffle between one Latin square and another. That is, see how extreme the outcome is in the original setting compared to how it would be if we shifted the dose labels, e.g., one shuffle might shift the labels to:

	east	**—**	**west**
north	Lo	Hi	Med
—	Med	Lo	Hi
south	Hi	Med	Lo

---

1.   This example is from Philip Good.

Terminology

Statistics has a lot of terminology. For the most part, we try to avoid it. On the other hand, you will hear other people use it, so you should know what it means.

acceptance region

The lower limit of the *p*-value at which point one accepts the **null hypothesis**. For example, if the cutoff is 5%, then one accepts the null hypothesis provided the *p*-value is 5% or greater. As Philip Good points out, academics commonly present the *p*-value without further interpretation. Industrial people have to decide. For example, suppose you are testing a drug and your null hypothesis is that it is non-toxic. You conduct a trial with 200 people, 100 of whom get the drug. If 23 people fall sick using the drug and only 21 when not, you'd probably reject the drug even though the *p*-value is quite high (and therefore there is no statistical reason to reject the null hypothesis that the drug is non-toxic). See **rejection region** and *p*-**value**.

alternate hypothesis

See **null hypothesis**.

blocking effect

If your independent variable is X and there is another possible confounding factor F and you test $X = x_1$ and $F = f_1$ and a second test $X = x_2$ and $F = f_2$, then the confounding factor F may swamp the effect of the independent variable.

block design

This is to avoid the **blocking effect** as follows: if F might be a confounding factor to the independent variable X, then make sure that each setting of X is tested with every value of F. Example: you want to test the wearability of a different shoe type so you have shoe types s_1 and s_2. Confounding factor is the person. So, if you have people $p_1, p_2, ..., p_n$, it would be bad to give s_1 to the first person and s_2 to the second one, and so on. Instead, give s_1 and s_2 to p_i with a random selection of left and right foot (another possible confounding factor). Thus for each s_i , you want to use all possible values of the confounding factor approximately equally. The **Latin square approach** quoted earlier is another technique to avoid a blocking effect.

distribution-free test

Is one whose result does not depend on any assumption about the underlying distribution. The **significance** test based on **shuffling** is distribution-free. The only assumption it makes is that if the **null hypothesis** did hold then one could rearrange the labels without affecting the underlying distribution of the outcomes. For such tests, the null hypothesis is usually exactly this (informally, the labels, e.g., treatments, don't matter). So this is not much of an assumption. The **bootstrap** is also distribution-free.

exact test

Is one that correctly assigns a p-value without regard to the distribution. See **distribution-free** and p-**value**.

non-parametric

See **distribution-free**.

null hypothesis

We have a null (boring) hypothesis (H_0) which is that whatever we see is due to chance alone, and an alternate (exciting) hypothesis that whatever we see is not due to chance alone, but something else. One hypothesis must be true and they cannot both be true. The question you have to decide, based on your data, is whether your null hypothesis is supported by your data, or if the null hypothesis should be rejected in favor of your exciting hypothesis.

one-tailed

See **tails**.

p-**value**

In a **significance** test the p-value is the probability of observing a summary from the data at least as extreme as the one calculated from the data under the assumption that the **null hypothesis** is true. For example, if we flip a supposedly fair coin 100 times, the p-value would be greater for the outcome 59 heads and 41 tails assuming the coin is fair than for the outcome 89 heads and 11 tails under the same assumption. This could be calculated by doing 10,000 bootstraps assuming a fair coin and counting the number of times the result was more extreme than this outcome. Call that number N. The p-value would then be N/10,000. (Of course, the binomial theorem could also be used.) Similarly, if the null hypothesis is that a treatment doesn't matter and a sample S shows that the treatment results differ by 30% from the placebo results, then the p-value of this null hypothesis can be computed by **shuffling** the labels (treatment/non-treatment) on S 10,000 times and counting how often the result is at least as extreme as the result observed. Call that number N. The p-value would then be N/10,000.

parametric method

Is one whose **significance** estimates depend on a specific distribution.

parametric bootstrap

Is **bootstrapping** according to a known distribution (e.g., the **fair coin assumption**) to see whether it is likely that some observed outcome is likely to have resulted from that distribution.

permutation test

Is what we are calling a **significance** or **shuffle** test.

power of a test

Is the probability that the test will reject the **null hypothesis** if the null hypothesis should be rejected. One can get a power of 1, if one always rejects the null hypothesis, but this fails to be *conservative* (i.e., one rejects too often). In general, the greater the **p-value** that separates the **rejection** from the **acceptance region**, the greater the power. The power also increases the larger the size of the sample. Given a desired power and either a distribution D reflecting your prior belief or an **empirical distribution**, you can determine the size of the sample you need to achieve that power as explained in the pragmatic consideration section (Chapter 2) above.

rank transformation

Is a **transformation** on a collection of numbers in which the numbers are sorted and then each number is replaced by its order in the sorted sequence. The sorting can be ascending or descending. You have to specify which.

rejection region

p-value below which you reject the **null hypothesis**. For example, the p-value that in 100 flips of a fair coin, there are 89 heads is so low that we would reject the hypothesis of fairness. Suppose we were conducting an experiment of the effectiveness of a drug and we conducted a trial of 200 people, 100 of whom received the drug and 100 not. The null hypothesis is that the drug is ineffective. If 23 of the 100 who received the drug got better but only 21 of the 100 who didn't receive the drug got better, then the p-value would be high, so there would be no reason to try the drug. See **acceptance region** and **p-value**.

tails

Suppose you are testing two treatments — yours and the competition. In a sample, yours seems to perform better. You want to see if the improvement is **significant**. The **null hypothesis** is that the other treatment is as good as yours or better. The **alternative** is that yours is better (a one-tailed alternative). So, you do a significance test. If it is very rare for the **shuffled** labels to show an advantage for your label as great or greater than the one you see, then you reject the null. Suppose now that you are a consumer watch organization and you are skeptical that there is any difference at all between the two treatments. You take a sample and you observe a difference with treatment A giving a higher value than treatment B. Your null hypothesis is that there is no difference. You reject the null if the probability that a shuffle gives as large a magnitude difference *in either direction* is small. This is called a two-tailed test.

transformation

Is any change to the data (e.g., taking the log of all values) that is called for either by the application or to use a certain test. In item #6 found in Ch. 3 above, we subtracted 5 centimeters from the height of each Dutch male in a sample to be able to test the null hypothesis that Dutch men were on average 5 centimeters taller than Italian men. Log transforms are useful when the difference between two treatments is expected to be multiplicative (e.g., a factor of 2) rather than additive.

two-tailed

See **tails**.

type I error

A type I error is thinking that the null hypothesis is false when it's in fact true (*false positive*). The *p*-value represents the probability of a type I error when you assert the null hypothesis is rejected.

type II error

A type II error is thinking that the null hypothesis is true when in fact it is not (*false negative*). Generally, if we reduce the probability of a type I error, we increase the probability of a type II error.

The Essential Stats

Our approach from now on is to name a statistic, discuss its domain of application, show how to calculate it, give a small example, give pseudocode for **confidence intervals** and **significance** as appropriate, and then a link to code. These can be read in any order.

Mean

A representative value for a group. This is what people are usually referring to when they use the term average.

Difference between Two Means

Used to compare two groups.

Chi-Squared

Usually used to measure the deviation of observed data from expectation or to test the independence of two variables.

Fisher's Exact Test

Used instead of Chi-Squared when one or more of the expected counts for the four possible categories ($2 \cdot 2 = 4$) are below ten.

One-way ANOVA

Used to measure how different two or more groups are when a single independent variable is changed.

Multi-way ANOVA

Used to test the influence of two or more factors (independent variables), on our outcome, when we have two or more groups.

Linear Regression

Used to find the line that best fits data so that it can be used to predict y given a new x.

Linear Correlation

Used to determine how well one variable can predict another (if a linear relationship exists between the two).

Multiple Regression

Used to determine how well a set of variables can predict another.

Multiple Testing

If a statistical test is run many times we expect to see a few values that look significant just due to chance. We discuss several ways to address this problem here.

4.1   Mean

4.1.1   Why and when

If you had just one way to characterize a collection, the mean would probably be the one to use. The mean also has a physical interpretation. If each number in a group is assigned unit weight and rests at its number position on a number line, then the center of gravity is the mean.

4.1.2   Calculate

Sum of all elements in a group/number of elements in the group.

4.1.3   Example

Say we have developed a new allergy medication and we want to determine the mean time to symptom relief. Here we record the time, in minutes, it took 10 patients to experience symptom relief after having taken the medication (if this were a real study, we would have far more patients): 60.2 63.1 58.4 58.9 61.2 67.0 61.0 59.7 58.2 59.8.

We want a range of values within which we are 90% confident that the true mean lies.

4.1.4   Pseudocode & code

All code was tested using Python 2.3.
www.python.com

download code and input files here:
http://www.morganclaypool.com/userimages/ContentEditor/1216676640846/Statistics_is_Easy_Codes.zip

Here is an example run of the MeanConf.py code:

```
Observed mean: 60.75
We have 90.0 % confidence that the true mean is between: 59.57 and 62.15
```

4.2   Difference between Two Means

4.2.1   Why and when

Very often someone may claim to improve something and give you before and after data, call them *B* and *A*. To evaluate the claim, you might calculate the mean of *B* and the mean of *A* and see if the latter is greater than the former (assuming higher values are better). But then two questions arise: (i) Could this improvement have arisen by chance? (ii) How much of a gain should we expect, expressed as a range?

4.2.2 Calculate

Compute the mean of *A* and the mean of *B* and find the difference.

4.2.3 Example

See Chapter 1.

4.2.4 Pseudocode & code

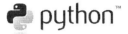

All code was tested using Python 2.3.
www.python.com

download code and input files here:
http://www.morganclaypool.com/userimages/ContentEditor/1216676640846/Statistics_is_Easy_Codes.zip

Here is an example run of the Diff2MeanSig.py code:

```
Observed difference of two means: 12.97

7 out of 10000 experiments had a difference of two means greater than or equal to
12.97 .

The chance of getting a difference of two means greater than or equal to 12.97 is
0.0007 .
```

Here is an example run of the Diff2MeanConf.py and Diff2MeanConfCorr.py code:

```
Observed difference between the means: 12.97

We have 90.0 % confidence that the true difference between the means is between:
7.81 and 18.11
```

4.3 Chi-squared

4.3.1 Why and when

The chi-squared statistic is usually used to measure the deviation of observed data from expectation or to test the independence of two group categories.

First we will use it to test if data has an expected distribution.

4.3.2 Calculate with example

We want to test to see if a die is fair. There are 6 possible outcomes (categories) when the die is rolled: 1, 2, 3, 4, 5, 6. Roll the die 60 times and keep track of the results in a frequency chart. We will also record the expected outcome for each category. We expect each category to have a probability of occurring of 1/6, so if we roll the die 60 times we expect each side to appear 10 times.

Used for *categorized data*.

Do n trials and make a frequency table that contains expected number of hits in each category i as well as the actual number of hits in each category i.

$$X^2 = \left(\frac{\left(X_1 - E(X_1)\right)^2}{E(X_1)} \right) + \dots + \left(\frac{\left(X_n - E(X_n)\right)^2}{E(X_n)} \right)$$

Category	X_i	$E(X_i)$	$X_i - E(X_i)$	$(X_i - E(X_i))^2$	$(X_i - E(X_i))^2 / E(X_i)$
Description	Observed number of times die landed on side i	Expected number of times die will land on side i	Difference of observed from expected	Difference of observed from expected squared	
1	14	10	4	16	1.6
2	16	10	6	4	3.6
3	6	10	-4	16	1.6
4	9	10	-1	1	.1
5	5	10	-5	25	2.5
6	10	10	0	0	0
Totals	60	60*	0**	-	$X^2 = 9.4$

* This total must always be the same as the observed total since it is the number of trials.
** This must always be zero, since if we observed more than we expected in one category, we must have observed less than we expected in others.

Rule of thumb: The expected number of outcomes in each category must be 10 or larger, and we must have at least 2 degrees of freedom.

$$df = n - 1$$

Where n is equal to the number of categories, in this case 6.

There are $n - 1$ degrees of freedom because once we know $X_1, X_2, \dots, X_{n-1}$, we know X_n. In our example we have 5 degrees of freedom, and each expected category count is at least 10, so we meet the requirements for a chi-squared test.

If the difference between observed counts and expected counts is large, we reject the hypothesis that the difference between the observed data and expected data is from chance alone, but how do we know when X^2 is large enough to do this?

We want to determine what the chances are, if the die really is fair, that we get a X^2 of 9.4 as we did above. If the die is fair would this happen rarely? Would it happen frequently? We figure this out by simulating 60 rolls of a fair die, 10,000 times. Draw 10,000 samples of size 60 and compute chi-squared statistic for each. See how many times the chi-squared statistic is greater than or equal to 9.4. This is our **p-value**, the chance that, given a fair die, we would get a chi-squared value at least as great as ours.

4.3.3   Pseudocode & code

All code was tested using Python 2.3.
www.python.com

download code and input files here:
http://www.morganclaypool.com/userimages/ContentEditor/1216676640846/Statistics_is_Easy_Codes.zip

When we ran the ChiSquaredOne.py code on the example above we got:

```
Observed chi-squared: 9.40
963 out of 10000 experiments had a chi-squared difference greater than or equal
to 9.40
Probability that chance alone gave us a chi-squared greater than or equal to 9.40
is 0.0963
```

4.3.4   Calculate with example for multiple variables

Here is another example.

Say we're looking at three income levels and sickness vs. health. Then we might have a table like this:

	Poor	**Middle**	**Rich**	**Total**
Sick	20	18	8	46
Expected	*(46 / 110) · 44 = 18.48*	*(46 / 110) · 42 = 17.64*	*(46 / 110) · 24 = 10.08*	
Healthy	24	24	16	64
Expected	*(64 / 110) · 44 = 25.52*	*(64 / 110) · 42 = 24.36*	*(64 / 110) · 24 = 13.92*	
Total	44	42	24	110

This example is a little more complicated than the previous one because we have two variables: wealth and health status. This affects how we calculate our expected values, how we calculate the degrees of freedom, and how we test for significance.

If we are asking ourselves how wealth affects health, wealth is our independent variable and health is our dependent variable. Our **null hypothesis** will be that changing wealth does not affect health. So we look at the ratio of total sick to total people, and we expect that ratio to hold across the whole sick row because wealth should not change it. In our example, we have 46 sick people, of the 110 total: 46/110 = 0.42. Now we look at each of the cells in the sick row and fill in the expected counts. We have a total of 44 poor people, so we *expect* 42% of them to be sick. 42% of 44 is 18.4 (obviously we can't have .4 of a person, but that is OK). We do this for all cells in the row. The same principle holds for the healthy row. There are 64 healthy people of the 110 people sampled—64/110 = 0.58—so across all wealth categories we *expect* 58% of the people to be healthy.

Category	X_i	$E(X_i)$	$X_i - E(X_i)$	$(X_i - E(X_i))^2$	$(X_i - E(X_i))^2 /$ $E(X_i)$
Description	Observed counts	Expected counts	Difference of observed from expected squared	Difference of observed from expected square	
Sick/Poor	20	18.48	1.52	2.31	.125
Sick/Middle	18	17.64	.36	.13	.001
Sick/Rich	8	10.08	-2.08	4.34	.43
Healthy/Poor	24	25.52	-1.52	2.31	.09
Healthy/Mid-dle	24	24.36	-.36	.13	.00053
Healthy/Rich	16	13.92	2.08	4,33	.31
Totals	110	110*	0**	-	$X^2 = .95653$

* This total must always be the same as the observed total since it is the number of trials.
** This must always be zero, since if we observed more than we expected in one category, we must have observed less than we expected in others.

Don't forget to check to make sure we can use the chi-squared test. We do have at least 10 for each expected category count, but we also need to make sure we have at least 2 degrees of freedom.

$$df = (r - 1)(c - 1)$$

where r is the number of rows and c is the number of columns. This is because once we know rows $1...r - 1$ and once we know columns $1...c - 1$, we know the values for the cells in row r and column c.

So in our example we have $(2 - 1) \cdot (3 - 1) = 2$ degrees of freedom.

In this example we test for significance a little differently than the last. Here we use the shuffle method. The shuffle must preserve the marginals (i.e., the totals for rows and the totals for columns).

The row marginals are: total sick is 46 and total healthy is 64. The column marginals are: total poor is 44, total middle is 46, and total rich is 24. That is, you compute a chi-square value for this. Now a shuffle is a rearrangement that keeps the marginals the same. For example, if you move one from sick/poor to sick/middle, you would not affect the sick marginal but you would affect the poor and middle marginals, so you might move 1 from healthy/middle to healthy/poor. Another way to think about his is that we invent $20 + 18 + 8 + 24 + 24 + 16 = 110$ individuals, each associated with a health label and a wealth label. So the first 20 would be poor and sick, the next 18 middle and sick, etc. Then we shuffle the wealth labels and reevaluate the chi-squared. Clearly, the marginals don't change.

We can't change the marginals because when we test for significance we depend on our expected values and our expected values are computed from the marginals. Remember when we test for significance we simulate taking 10,000 samples — we use the expected probabilities to generate these samples.

4.3.5   Pseudocode & code

All code was tested using Python 2.3.
www.python.com

download code and input files here:
http://www.morganclaypool.com/userimages/ContentEditor/1216676640846/Statistics_is_Easy_Codes.zip

When we ran the ChiSquaredMulti.py code on the example above we got:

```
Observed chi-squared: 0.97
6241 out of 10000 experiments had a chi-squared greater than or equal to 0.97
Probability that chance alone gave us a chi-squared greater than or equal to 0.97
is 0.6241
```

Based on these results we conclude that wealth has no significant affect on health (remember that our observed values are made up!).

4.4   Fisher's Exact Test

4.4.1   Why and when

Fisher's exact test can be used instead of **chi-squared** when you have two variables (for example health and wealth), each having two categories (for example: sick, healthy and poor, rich), and one or more of the expected counts for the four possible categories ($2 \cdot 2 = 4$) are below 10 (remember: you cannot use chi-squared if even one expected count is less than 10). Like chi-squared, Fisher's exact test can be used to see if there is a relationship between the two variables as well as to measure the deviation of observed data from expectation.

Reputedly, Fisher was inspired to develop this test after an acquaintance, Dr. Muriel Bristol, claimed to be able to distinguish between a cup of tea in which the milk was poured before the tea and a cup in which the tea was poured before the milk[1]. In honor of Dr. Bristol, we will test this hypothesis.

---

1.   http://en.wikipedia.org/wiki/Muriel_Bristol

4.4.2 Calculate with Example

Say we have collected the following data:

		Actual Order		
		Milk First	Tea First	Total
Claim of Tea Taster	Milk First	3	1	4
	Tea First	2	4	6
	Total	5	5	10

The above generalizes to:

		Variable A		
		Category A1	Category A2	Total
Variable B	Category B1	a	b	$a + b$
	Category B2	c	d	$c + d$
	Total	$a + c$	$b + d$	n

If our tea taster really can correctly identify which cups of tea had the milk poured first and which did not, then we expect a and d to be large (meaning he or she guessed correctly most of the time) and b and c to be small (incorrect guesses). If b and c are large, then our taster consistently misidentified the tea. If our taster consistently misidentifies the tea then he or she probably can distinguish between the two types, but has labeled them incorrectly. If our taster cannot distinguish between types of tea, then we expect $a, b, c,$ and d to all be about the same, meaning the taster was wrong about as often as was correct.

Our **null hypothesis** is that our tea taster will not be able to tell the difference between the two types of teas. Notice here we are stating that we do not think that our taster can distinguish between the teas, we are *not* claiming that he or she cannot correctly identify them. So if our taster misidentified every single cup of tea, we would probably conclude that he or she can distinguish between the two types of tea. These are two different hypotheses and we will address the latter one later.

This is called an *exact* test because we know all of the possible 2 x 2 matrices we could have gotten while preserving the marginals (i.e., keeping the same row and column totals), and we know the exact probability of getting each matrix by chance given our **null hypothesis** is true. Remember that our ***p*-value** always includes the probability that chance alone gave us our observed data as well as the probability of chance alone giving us data even more *extreme* than ours. Being very clear about what question you are asking (what your null hypothesis is) will help you identify which cases are at least as extreme as your observed case.

The probability of getting matrix

a	b
c	d

is:

$$\frac{(a+b)!(c+d)!(a+c)!(b+d)!}{a!b!c!d!n!}$$

Remember that:

$$x! = x \cdot (x - 1) \cdot (x - 2) \cdot \ldots \cdot 1$$

And that $0! = 1$.

We know that the sum of the probabilities of all possible outcomes must be 1. So the denominator in the formula above represents the whole (i.e., all possible outcomes). The numerator in the above formula represents the number of times we expect to get this particular outcome. A large numerator yields a large probability, meaning this outcome is quite likely to occur. The numerator will be large when the matrix is balanced (i.e., a, b, c, and d are about equal). The more unbalanced the matrix is, the smaller the numerator will be. Bringing this back to our original tea problem, the greater the difference is between the number of teas our taster correctly identified (a and d) and the number he or she incorrectly identified (b and c), the smaller the likelihood that this difference is due to chance alone. This makes sense, if the taster gets them all mostly right, or all mostly wrong, it suggests that there is in fact a relationship of some kind between his or her decision making process, and the actual kind of tea he or she was given. If there was no relationship between the two variables (claim of the taster and actual order of milk and tea) then we expect our taster to be wrong about the same number of times as he or she is correct.

Here are all of the matrices we could have gotten given our row and column totals, including the one we did get, and their associated probabilities:

Matrix			Probability Calculation	Probability
0	4	**4**	$((0 + 4)!(5 + 1)!(0 + 5)!(4 + 1)!) / (0!4!5!1!10!)$	
5	1	**6**	$= (4!6!5!5!) / (0!4!5!1!10!)$	
5	**5**	**10**	$= (6!5!) / (10!)$	
(1 correct, 9 incorrect)			$= (6 \cdot 5 \cdot 4 \cdot 3 \cdot 2 \cdot 1 \cdot 5 \cdot 4 \cdot 3 \cdot 2 \cdot 1) / (1 \cdot 10 \cdot 9 \cdot 8 \cdot 7 \cdot 6 \cdot 5 \cdot 4 \cdot 3 \cdot 2 \cdot 1)$	0.0238
			$= (5 \cdot 4 \cdot 3 \cdot 2) / (10 \cdot 9 \cdot 8 \cdot 7)$	
			$= (4 \cdot 3) / (9 \cdot 8 \cdot 7)$	
			$= 1 / (3 \cdot 2 \cdot 7)$	
			$= 1 / 42$	

1	3	4	$((1 + 3)!(4 + 2)!(1 + 4)!(3 + 2)!) / (1!3!4!2!10!)$	
4	2	6	$= (4!6!5!5!) / (1!3!4!2!10!)$	
5	5	10	$= (6 \cdot 5 \cdot 4 \cdot 3 \cdot 2 \cdot 1 \cdot 5 \cdot 4 \cdot 3 \cdot 2 \cdot 1 \cdot 5 \cdot 4 \cdot 3 \cdot 2 \cdot 1) / (3 \cdot 2 \cdot 1 \cdot 2 \cdot 1 \cdot 10 \cdot 9 \cdot 8 \cdot 7 \cdot 6 \cdot 5 \cdot 4 \cdot 3 \cdot 2 \cdot 1)$	0.2381
(3 correct, 7 incorrect)			$= (5 \cdot 4 \cdot 5 \cdot 4 \cdot 3) / (10 \cdot 9 \cdot 8 \cdot 7)$	
			$= 5 / 21$	
			$= 10 / 42$	
2	2	4	$((2 + 2)!(3 + 3)!(2 + 3)!(2 + 3)!) / (2!2!3!3!10!)$	
3	3	6	$= (4!6!5!5!) / (2!2!3!3!10!)$	
5	5	10	$= (4 \cdot 3 \cdot 2 \cdot 1 \cdot 6 \cdot 5 \cdot 4 \cdot 3 \cdot 2 \cdot 1 \cdot 5 \cdot 4 \cdot 3 \cdot 2 \cdot 1 \cdot 5 \cdot 4 \cdot 3 \cdot 2 \cdot 1) / (2 \cdot 1 \cdot 2 \cdot 1 \cdot 3 \cdot 2 \cdot 1 \cdot 3 \cdot 2 \cdot 1 \cdot 10 \cdot 9 \cdot 8 \cdot 7 \cdot 6 \cdot 5 \cdot 4 \cdot 3 \cdot 2 \cdot 1)$	0.4762
(5 correct, 5 incorrect)			$= 20 / 42$	
observed frequency			$((3 + 1)!(2 + 4)!(3 + 2)!(1 + 4)!) / (3!1!2!4!10!)$	
3	1	4	$= 10 / 42$ (as above)	
2	4	6		0.2381
5	5	10		
(7 correct, 3 incorrect)				
4	0	4	$((4 + 0)!(1 + 5)!(4 + 1)!(0 + 5)!) / (4!0!1!5!10!)$	
1	5	6	$= 1 / 42$ (as above)	
5	5	10		0.0238
(9 correct, 1 incorrect)				
The sum of all probabilities must be 1.				

Remember we are interested in matrices that are at least as extreme as the observed matrix. Our taster correctly identified 7 cups of tea and incorrectly identified 3. We decided that our **null hypothesis** is that our taster cannot distinguish between the two types of tea, which means that cases that are as extreme, or more extreme than our observed case, are those in which our taster got 3 or fewer incorrect, as well as those in which our taster got 3 or fewer correct. The first, second, fourth, and fifth matrices make up the set of matrices at least as extreme as the observed matrix (they have a probability that is less than or equal to our observed matrix). The sum of these probabilities gives us our *p*-value.

$$p\text{-value} = 0.0238 + 0.2381 + 0.2381 + 0.0238 = 0.5238$$

Assuming the **null hypothesis** is true and our tea taster cannot distinguish between cups of tea that had the milk poured in before the tea and those that did not, we would expect to get observed values such as ours 52% of the time. If the null hypothesis is true it is quite likely we would see results like ours, so we should not reject the null hypothesis; most likely our tea taster cannot differentiate between the two types of tea. Of course this does not mean that Dr. Bristol could not distinguish between the two, perhaps she had a more discriminating tea palate.

The above is an example of a two-**tailed** test. Now if our **null hypothesis** had been that our taster cannot correctly identify the tea cups, our definition of more extreme would only include cases in which our taster got even more correct than in our observed case. There is only one matrix in which that could have occurred: if she had gotten 9 correct and 1 incorrect. This is a one-**tailed** test. Our ***p*-value** here is the sum of the probability of getting matrix 4 and the probability of getting matrix 5.

$$p\text{-value} = 0.2381 + 0.0238 = 0.2619$$

So assuming our taster cannot correctly identify the cups of tea, we would expect to see an outcome such as ours about 26% of the time. This is still too probable for us to reject this hypothesis.

4.4.3 Pseudocode & code

All code was tested using Python 2.3.
www.python.com

download code and input files here:
http://www.morganclaypool.com/userimages/ContentEditor/1216676640846/Statistics_is_Easy_Codes.zip

Here is an example run of the FishersExactTestSig.py code:

```
Observed Fisher's Exact Test: 0.2619
3997 out of 10000 experiments had a Fisher's Exact Test less than or equal to
0.2619
Probability that chance alone gave us a Fisher's Exact Test of 0.2619 or less is
0.3997
```

4.5 One-Way ANOVA

4.5.1 Why and when

Given two or more groups, how would we measure how different they are? The mean of each group is a reasonable statistic to compare the groups. R. A. Fisher developed a statistic, called the *f*-statistic in his honor, for exactly this purpose.

4.5.2 Calculate with example

The one-way ANOVA, or single-factor ANOVA, tests the effect of a single independent variable on the groups. The groups must be categorized on this single variable. The idea is that we want to compare the variance between the groups with the variance within the groups. If the variance between the groups is significantly greater than the variance within the groups, we conclude that the independent variable does affect the groups.

High variance within groups; little variance between groups.

Zero variance within groups; large variance between groups.

Figure 4.1:   The difference between within group variance and between group variance.

Here we have an exaggerated example of the difference between within group variance and between group variance. In the top row we have three groups with large within group variance, but that are exactly the same so there is no between group variance. In the bottom row we have three groups that have no within group variance, but the groups are different from each other, and therefore have a large between group variance.

Note: The number of elements in each group need not be equal.

How to compute the *f*-statistic:

This algorithm involves a lot of steps so we will explain it using a simple example and a table. We have three different drugs used as treatment. We want to test the effect of our independent variable, the drug given as treatment, on the number of days it takes a patient to fully recover, so we separate patients into three groups based on the drug they have been given and record the number of days it takes for each to recover:

$$\text{Drug A} = 45, 44, 34, 33, 45, 46, 34$$
$$\text{Drug B} = 34, 34, 50, 49, 48, 39, 45$$
$$\text{Drug C} = 24, 34, 23, 25, 36, 28, 33, 29$$

We compare the between group variance and the within group variance using the ratio:

$$MS_{Between}/MS_{Within}$$

where $MS_{Between}$ is the between group variance, referred to as the between group mean squares and MS_{Within} is the within group variance, referred to as the within group mean squares.

To compute the within group mean squares we compute the within group sum of squares and divide it by the within group degrees of freedom. The within group sum of squares will be the sum of each group's sum of squares. This may sound confusing, but it is actually straightforward. To get one group's sum of squares we compute the difference between each value in the group and that value's own group mean, square the result, and then sum these squares.

$$W_{SS} = \sum_{g=1}^{k} \sum_{i=1}^{n_g} (x_i - \bar{X}_g)^2$$

where k is the number of groups, n_g is the number of elements in group g, x_i is the ith value in the group g, and $\bar{X}_g$ is the group mean.

Here we compute the sum of squares for each group in our example:

$$\text{Drug A}_{SS} = (45 - 40.14)^2 + (44 - 40.14)^2 + (34 - 40.14)^2 + (33 - 40.14)^2 + (45 - 40.14)^2 + (46 - 40.14)^2 + (34 - 40.14)^2 = 222.86$$

$$\text{Drug B}_{SS} = (34 - 42.71)^2 + (34 - 42.71)^2 + (50 - 42.71)^2 + (49 - 42.71)^2 + (48 - 42.71)^2 + (39 - 42.71)^2 + (45 - 42.71)^2 = 291.43$$

$$\text{Drug C}_{SS} = (24 - 29)^2 + (34 - 29)^2 + (23 - 29)^2 + (25 - 29)^2 + (36 - 29)^2 + (28 - 29)^2 + (33 - 29)^2 + (29 - 29)^2 = 168$$

The within group sum of squares equals 682.29; it is the sum of the above group sum of squares.

The last piece of information we need to compute the within group mean squares (the within group variance) is the degrees of freedom for the within group sum of squares. In general,

$$df = n - 1$$

where n is equal to the number of values in the group.

The degrees of freedom measures the number of independent pieces of information on which our statistic is based. When computing the group df we subtract one from the group count because if we know the mean and n, once we know the $(n - 1)^{th}$ values we know what the nth value must be.

Therefore, the degrees of freedom for the within group sum of squares is:

$$W_{df} = N - k$$

where N is equal to the total number of values in all the groups, and k is equal to the number of groups. We subtract k from N because for each group we lose one degree of freedom.

Computing the between group mean squares is very similar to computing the within group mean squares, except that for each value we get the difference between the value's own group mean and the total mean (computed by pooling all the values).

$$B_{SS} = \sum_{i=1}^{N} (\bar{X}_g - \bar{X})^2$$

Where N is equal to the total number of values in all the groups, $\bar{X}_g$ is the group mean of the ith value in the pool of all values, and is the overall mean. Because every value in a particular group will have the same (group mean – total mean), the above formula can be simplified to:

$$B_{SS} = \sum_{g=1}^{k} n_g (\bar{X}_g - \bar{X})^2$$

Where k is equal to the number of groups, n_g is equal to the total number of values in group g, $\bar{X}_g$ is the group mean and $\bar{X}$ is the overall mean. That is, B_{SS} multiplies the squared difference of the group mean and total mean by the number of values in the group.

In our example:

$$B_{SS} = 7 \cdot (40.14 - 36.91)^2 + 7 \cdot (42.71 - 36.91)^2 + 8 \cdot (29 - 36.91)^2 = 812.65$$

The between group degrees of freedom is:

$$B_{df} = k - 1$$

where k is equal to the number of groups. We subtract 1 from k because if we know the number of elements in each group and the total group mean, once we know the $(k - 1)^{th}$ group means, we know what the kth group mean must be.

Group	Description	A	B	C
Count	Number of values in group	7	7	8
$\bar{X}_g$	Within group mean, where g is the group number	40.14	42.71	29.0
G_{ss}	Sum of squares for each group	222.86	291.43	168
W_{SS}	Within group sum of squares	682.29		

G_{df}	Group degrees of freedom	6	6	7
W_{df}	Within group degrees of freedom	19		
$\overline{X}$	Total mean, the mean of all the values pooled	36.91		
B_{SS}	Between group sum of squares	812.65		
B_{df}	Between group degrees of freedom	2		
T_{SS}	Total sum of squares	1491.81		
MS_W	Within group variance (mean squares)	35.91		
MS_B	Between group variance (mean squares)	406.33		
f	Measure of how much bigger the between group variance is than the within group variance	11.31		

The total sum of squares measures the variance of all the values. $T_{SS} = W_{SS} + B_{SS}$. Therefore $B_{SS} = T_{SS} - W_{SS}$ and $W_{SS} = T_{SS} - B_{SS}$ hold. If it is easier for you to compute W_{SS} and T_{SS}, you can use these to compute B_{SS}.

$$T_{SS} = \sum_{i=1}^{N} (x_i - \overline{X})^2$$

where N is equal to the total number of values in all the groups, x_i is the ith value in the pool of all values, and $\overline{X}$ is the overall mean.

We now have our f-statistic, 11.31, so we know that the between group variance is greater than the within group variance, but we still do not know whether or not this difference is **significant**. We use the shuffle method to compute the significance of our f-statistic.

4.5.3 Pseudocode & code

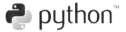

All code was tested using Python 2.3.
www.python.com

download code and input files here:
http://www.morganclaypool.com/userimages/ContentEditor/1216676640846/Statistics_is_Easy_Codes.zip

When we ran the OneWayAnovaSig.py code on the example above we got:

```
Observed F-statistic: 11.27
10 out of 10000 experiments had a F-statistic difference greater than or equal to
11.27
Probability that chance alone gave us a F-statistic of 11.27 or more is 0.001
```

OK, the difference may be significant, but is it large? Does it matter?

All code was tested using Python 2.3.
www.python.com
download code and input files here:
http://www.morganclaypool.com/userimages/ContentEditor/1216676640846/Statistics_is_Easy_Codes.zip

When we ran the OneWayAnovaConf.py code we got:

```
Observed F-statistic: 11.27
We have 90.0 % confidence that the true F-statistic is between: 6.28 and 29.60
***** Bias Corrected Confidence Interval *****
We have 90.0 % confidence that the true F-statistic is between: 4.47 and 20.35
```

All f tells you is whether or not there is a **significant** difference between the group means, not which group means are significantly different from each other. You can do follow-on tests to find out which group means are different since we have decided our f-statistic indicates the group means significantly differ. You can now compare the different groups by **comparing the group means**. You could do this pairwise comparison between every group, or if you have a control group you can do pairwise comparisons between the control and every other group.

4.6 Multi-way ANOVA

4.6.1 Why and when

We use a multi-way ANOVA, or multi-factor ANOVA, when we want to test the influence of two or more *factors (independent variables)*, on our outcome. Each factor must be discrete, or categorical. For example, this is not the appropriate test to ask if sex and weight effect some measure of health, unless weight is divided into categories such as underweight, healthy weight, etc.

In our **one-way ANOVA** example, we tested the effect of only one factor—the drug a patient was given—on outcome. We used this factor to split our outcome measures into different groups, and then used the f-statistic to decide whether or not our groups were significantly different. The steps we take for the multi-way ANOVA are very similar to the **one-way ANOVA**, but with each additional factor we add additional sources of variance, and we must take these additional sources of variance into consideration.

4.6.2   Calculate with example

For demonstrative purposes we will do a two-factor ANOVA comparing drug treatment and sex to outcome.

Here is a table that summarizes our data.

		factor S		Totals
		Category s_1	Category s_2	
factor D	Category d_1	15	13	$\overline{X}_{d_1} = 13.13$
		12	13	
		13	12	
		16	11	
		$\overline{X}_{d_1 s_1} = 14$	$\overline{X}_{d_1 s_2} = 12.25$	
	Category d_2	19	13	$\overline{X}_{d_2} = 14.88$
		17	11	
		16	11	
		15	17	
		$\overline{X}_{d_2 s_1} = 16.75$	$\overline{X}_{d_2 s_2} = 13$	
	Category d_3	14	11	$\overline{X}_{d_3} = 12.71$
		13	12	
		12	10	
		17	$\overline{X}_{d_2 s_3} = 11$	
		$\overline{X}_{d_3 s_1} = 14$		
Totals		$\overline{X}_{s_1} = 14.92$	$\overline{X}_{s_2} = 12.18$	$\overline{X} = 13.6$

We have split our outcome data into six groups. In a multi-way ANOVA if we have F factors, we end up with

$$\prod_{f=1}^{F} n_f$$

groups, where n_f is the number of categories in factor f. In our example, we have factor D, drug, of which there are 3 possibilities, and factor S, sex, of which there are 2 possibilities. Since there are 3 categories in factor D and 2 in factor S, we end up with $3 \cdot 2 = 6$ groups.

In the **one-way ANOVA** we compared the between group variance to the within group variance to see if there was a difference in the means of our groups. In the **one-way ANOVA** we only had one factor that could have been introducing variance outside of natural variance between the participants in the study. In a two-factor ANOVA, we have:

1. the natural variance between the study participants
2. the variance in the groups caused by factor D
3. the variance in the groups caused by factor S
4. the variance in the groups caused by the *interaction* of factor D and factor S.

Using a two-factor ANOVA we can find out:

1. how factor D influences the group means
2. how factor S influences the group means
3. how the *interaction* of factor D and factor S influences the group means

Remember: The sum of squares is the sum of squared differences of each value from some mean. We use the sum of squares and the degrees of freedom to measure variance in a group or between groups. So here we will use this to measure all of the sources of variance.

We compute the within group variation just as we did in the **one-way ANOVA**.

$$W_{SS} = \sum_{g=1}^{k} \sum_{i=1}^{n_g} (x_i - \overline{X}_g)^2$$

where k is the number of groups, n_g is the number of elements in group g, x_i is the ith value in the group g, and $\overline{X}_g$ is the group mean.

Here we compute the sum of squares for each group in our example:

Group $d_1 s_{1SS} = (15 - 14)^2 + (12 - 14)^2 + (13 - 14)^2 + (16 - 14)^2 = 10$

Group $d_1 s_{2SS} = (13 - 12.25)^2 + (13 - 12.25)^2 + (12 - 12.25)^2 + (11 - 12.25)^2 = 2.75$

Group $d_2 s_{1SS} = (19 - 16.75)^2 + (17 - 16.75)^2 + (16 - 16.75)^2 + (15 - 16.75)^2 = 8.75$

Group $d_2 s_{2SS} = (13 - 13)^2 + (11 - 13)^2 + (11 - 13)^2 + (17 - 13)^2 = 24$

Group $d_3 s_{1SS} = (14 - 14)^2 + (13 - 14)^2 + (12 - 14)^2 + (17 - 14)^2 = 14$

Group $d_3 s_{2SS} = (11 - 11)^2 + (12 - 11)^2 + (10 - 11)^2 = 2$

Within Group Sum of Squares = 10 + 2.75 + 8.75 + 24 + 14 + 2 = 61.5

The degrees of freedom for the within group sum of squares is:

$$W_{df} = N - k$$

where N is equal to the total number of values in all the groups, and k is equal to the number of groups. We subtract k from N because for each group we lose one degree of freedom.

In our example, the degrees of freedom for the within group sum of squares is 17 because we have 23 values in total and 6 groups.

Now we can compute the within group mean squares, which is the within group sum of squares divided by the within group degrees of freedom.

$$W_{MS} = 61.5 / 17 = 3.62$$

The between group sum of squares will include the variance from 3 sources: factor D, factor S, and the interaction of factors D and S. So we will refer to the between group sum of squares as the total between group sum of squares, to indicate that it incorporates all of these sources of variance. It is computed just as our between group sum of squares in the **one-way ANOVA**.

$$B_{SS} = \sum_{i=1}^{N} (\bar{X}_g - \bar{X})^2$$

where N is equal to the total number of values in all the groups, $\bar{X}_g$ is the group mean of the ith value in the pool of all values, and $\bar{X}$ is the overall mean.

In our example the total between group sum of squares equals:

$$B_{SS} = 4 \cdot (14 - 13.6)^2 + 4 \cdot (12.25 - 13.6)^2 + 4 \cdot (16.75 - 13.6)^2 + 4 \cdot (13 - 13.6)^2 + 4 \cdot (14 - 13.6)^2 + 4 \cdot (11 - 13.6)^2 = 69.98$$

The between group degrees of freedom is:

$$B_{df} = k - 1$$

where k is equal to the number of groups. We subtract 1 from k because if we know the number of elements in each group and the total group mean, once we know the $(k - 1)^{th}$ group means, we know what the kth group mean must be.

In our example, the between group degrees of freedom is 5, because we have 6 groups.

Now we want to split B_{SS} into its 3 parts: the main effects of factor D, the main effects of factor S, and the interaction effect.

To figure out how factor D influences the group means, we just do a **one-way ANOVA** where our groups are $d_1, d_2,$ and d_3.

$$\text{factorD}_{SS} = \sum_{i=1}^{N} (\bar{X}_d - \bar{X})^2$$

where N is equal to the total number of values in all the groups, i is the current value, d is the current group of factor D, $\bar{X}_d$ is the mean of the values in group d, and $\bar{X}$ is the overall mean. Because every value in a particular group d will have the same (group mean - total mean), the above formula can be simplified to:

$$\text{factorD}_{SS} = \sum_{d=1}^{D} n(\bar{X}_d - \bar{X})^2$$

where D is equal to the number of groups, or categories, in factor D, d is the current group of factor D, n is equal to the total number of values in group d, $\bar{X}_d$ is the mean of the values in group d, and $\bar{X}$ is the overall mean. We just multiplied the squared difference of the group mean and total mean by the number of values in the group.

In our example:

$$\text{factor D}_{SS} = 8 \cdot (13.13 - 13.6)^2 + 8 \cdot (14.88 - 13.6)^2 + 7 \cdot (12.71 - 13.6)^2 = 42.92$$

$$\text{factorD}_{df} = D - 1$$

In our example factor D has 2 degrees of freedom.

Now we can compute factor D mean squares, which is the factor D sum of squares divided by the factor D degrees of freedom.

$$\text{factor D}_{MS} = 42.92 / 2 = 21.46$$

We compute how factor S influences the group means in the same way.

$$\text{factorS}_{SS} = \sum_{i=1}^{N} (\bar{X}_s - \bar{X})^2$$

OR

$$\text{factorS}_{SS} = \sum_{s=1}^{S} n(\bar{X}_s - \bar{X})^2$$

In our example:

$$\text{factor S}_{SS} = 12 \cdot (14.92 - 13.6)^2 + 11 \cdot (12.18 - 13.6)^2 = 20.3$$

and

$$\text{factorS}_{df} = S - 1$$

In our example factor S has 1 degree of freedom.
Now we can compute factor S mean squares, which is the factor S sum of squares divided by the factor S degrees of freedom.

$$\text{factor } S_{MS} = 20.3 / 1 = 20.3$$

Finally,

$$\text{interaction}_{SS} = B_{SS} - \text{factorD}_{SS} - \text{factorS}_{SS}$$

In our example:

$$\text{interaction}_{SS} = 69.98 - 42.92 - 20.3 = 6.76$$

and

$$\text{interaction}_{df} = B_{df} - \text{factorD}_{df} - \text{factorS}_{df}$$
$$\text{interaction}_{df} = 5 - 2 - 1 = 2$$

Now we can compute the interaction mean squares, which is the interaction sum of squares divided by the interaction degrees of freedom.

$$\text{interaction}_{MS} = 6.76 / 2 = 3.38$$

Remember from the **one-way ANOVA** that our f-statistic compares the between group variance and the within group variance. The same holds with the multi-way ANOVA, except we will compare each part of the between group variance with the within group variance, to answer each of three hypotheses. So instead of:

$$\frac{MS_{Between}}{MS_{Within}}$$

we will compare factor D mean squares with the within group mean squares, factor S mean squares with the within group mean squares, and the interaction mean squares with the within group mean squares. In other words, we are comparing the variance caused by factor D with the variance within the groups, the variance caused by factor S with the variance within the groups, and the variance caused by the interaction of factors D and S to the variance within the groups.

$$\frac{MS_{factorD}}{MS_{Within}}, \quad \frac{MS_{factorS}}{MS_{Within}} \quad \text{and} \quad \frac{MS_{interaction}}{MS_{Within}}$$

Here we summarize our results:

	SS	df	MS	f
Within	61.5	17	3.62	
Between	69.98	5		
factor D	42.92	2	21.46	21.46/3.62 = 5.93
factor S	20.3	1	20.3	20.3/3.62 = 5.61
Interaction of D and S	6.73	2	3.38	3.38/3.62 = 0.93

We still don't know if this is significant.

4.6.3 Pseudocode & code

All code was tested using Python 2.3.
www.python.com

download code and input files here:
http://www.morganclaypool.com/userimages/ContentEditor/1216676640846/Statistics_is_Easy_Codes.zip

When we ran the TwoWayAnovaSig.py code on the example above we got:

```
Observed F-statistic: 0.93
4152 out of 10000 experiments had a F-statistic greater than or equal to 0.93
Probability that chance alone gave us a F-statistic of 0.93 or more is 0.4152
```

OK, the difference *is not* **significant**. If it were we would check if it is large; if the difference matters.

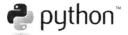

All code was tested using Python 2.3.
www.python.com

download code and input files here:
http://www.morganclaypool.com/userimages/ContentEditor/1216676640846/Statistics_is_Easy_Codes.zip

When we ran the TwoWayAnovaConf.py code we got:

```
Observed F-statistic: 0.93
We have 90.0 % confidence that the true F-statistic is between: 0.50 and 8.44
***** Bias Corrected Confidence Interval *****
We have 90.0 % confidence that the true F-statistic is between: 0.20 and 2.04
```

4.7 Linear Regression

4.7.1 Why and when

Linear regression is closely tied to **linear correlation**. Here we try to find the line that best fits our data so that we can use it to predict y given a new x. Correlation measures how tightly our data fits that line, and therefore how good we expect our prediction to be.

4.7.2 Calculate with example

> BE CAREFUL
> Regression can be misleading when there are outliers or a nonlinear relationship.

The first step is to draw a scatter plot of your data. Traditionally, the independent variable is placed along the x-axis, and the dependent variable is placed along the y-axis. The dependent variable is the one that we expect to change when the independent one changes. Note whether the data looks as if it lies approximately along a straight line. If it has some other pattern, for example, a curve, then a linear regression should not be used.

The most common method used to find a regression line is the least squares method. Remember that the equation of a line is:

$$y = bx + a$$

where b equals the slope of the line and a is where the line crosses the y-axis.

The least squares method minimizes the vertical distances between our data points (the observed values) and our line (the predicted values).

The equation of the line we are trying to find is:

$$y' = bx + a$$

where y' is the predicted value of y for some x.

We have to calculate b and a.

$$b = \frac{XY_{SP}}{X_{SS}}$$

where XY_{SP} is the sum of products:

$$XY_{SP} = \sum_{i=1}^{N}(x_i - \bar{X})(y_i - \bar{Y})$$

and X_{SS} is the sum of squares for X:

$$X_{SS} = \sum_{i=1}^{N} (x_i - \bar{X})^2$$

In our example:

	x_i	y_i	$\bar{X}$	$\bar{Y}$	$(x_i - \bar{X})$	$(y_i - \bar{Y})$	$(x_i - \bar{X})(y_i - \bar{Y})$	$(x_i - \bar{X})^2$
	1350	3.6	1353	3.5	-3	.1	-.3	9
	1510	3.8	1353	3.5	157	.3	47.1	24649
	1420	3.7	1353	3.5	67	.2	13.4	4489
	1210	3.3	1353	3.5	-143	-.2	28.6	20449
	1250	3.9	1353	3.5	-103	.4	-41.2	10609
	1300	3.4	1353	3.5	-53	-.1	5.3	2809
	1580	3.8	1353	3.5	227	.3	68.1	51529
	1310	3.7	1353	3.5	-43	.2	-8.6	1849
	1290	3.5	1353	3.5	-63	0	0	3969
	1320	3.4	1353	3.5	-33	-.1	3.3	1089
	1490	3.8	1353	3.5	137	.3	41.1	18769
	1200	3.0	1353	3.5	-153	-.5	76.5	23409
	1360	3.1	1353	3.5	7	-.4	-2.8	49
Totals	-	-	-	-	-	-	230.5	163677

$XY_{SP} = 230.5$ and $X_{SS} = 163677$ and $230.5 / 163677 = 0.0014$, so $b = 0.0014$.

The regression line will always pass through the point $(\bar{X}, \bar{Y})$ so we can plug this point into our equation to get a, where the line passes through the y-axis.

So far we have: $y' = 0.0014x + a$, so to solve for a we rearrange and get: $a = y' - 0.0014x$. Plug in (1353, 3.5) for (y', x) and we get $a = 3.5 - (0.0014 \cdot 1353) = 1.6058$. Our final regression equation is:

$$y' = .0014x + 1.6058$$

We **shuffle** the y values to break any relationship that exists between x and y. We do this 10,000 times, computing the slope each time to get the probability of a slope greater than or equal to ours given no relationship between x and y. If the slope was negative we would test to see the probability of getting a slope less than or equal to ours given no relationship between x and y. Remember that if the slope is zero, then x is not related to an increase or decrease in y.

4.7.3 Pseudocode & code

All code was tested using Python 2.3.
www.python.com

download code and input files here:
http://www.morganclaypool.com/userimages/ContentEditor/1216676640846/Statistics_is_Easy_Codes.zip

When we ran the RegressionConf.py code on the example above we got:

```
Line of best fit for observed data:  y' = 0.0014 x +  1.6333
165 out of 10000 experiments had a slope greater than or equal to 0.0014 .
The chance of getting a slope greater than or equal to 0.0014 is 0.0165 .
```

OK, the difference may be significant, but is it large? Does it matter?

When we ran the RegressionConf.py code on the example above we got:

```
Line of best fit for observed data:  y' = 0.0014 x +  1.6333
We have 90.0 % confidence that the true slope is between: 0.0005 and 0.0022
***** Bias Corrected Confidence Interval *****
We have 90.0 % confidence that the true slope is between: 0.0004 and 0.0022
```

4.8   Linear Correlation

4.8.1   Why and when

The correlation coefficient is a measure of strength and direction of a linear relationship between two random variables. Remember, we used **regression** to define the linear relationship between the two variables, and now we want to know how well X predicts Y.

4.8.2   Calculate & example

Draw scatter plot of data. Place the independent variable along the x-axis, and the dependent variable along the y-axis. We want to see what happens to the dependent variable when the independent one changes. Note whether the data looks as if it lies approximately along a straight line. If it has some other pattern, for example, a curve, then the linear correlation coefficient should not be used.

The correlation coefficient is always between -1 and 1. Values close to 1 indicate a strong positive association, meaning that as X increases we expect Y to increase (positive sloping line). Values close to -1 indicate a strong negative association, meaning that as X increases we expect Y to decrease (negative sloping line). A value of zero indicates that there is no relationship between the variables, that is, that knowing X does not help you predict Y.

The correlation coefficient, r, is computed by comparing the observed covariance (a measure of how much X and Y vary together) to the maximum possible positive covariance of X and Y.:

$$r = \frac{XY_{SP}}{\sqrt{X_{SS}Y_{SS}}}$$

where XY_{SP} is the sum of products, X_{SS} is the sum of squares for X, and Y_{SS} is the sum of squares for Y:

$$XY_{SP} = \sum_{i=1}^{N}(x_i - \bar{X})(y_i - \bar{Y})$$

$$X_{SS} = \sum_{i=1}^{N}(x_i - \bar{X})^2$$

$$Y_{SS} = \sum_{i=1}^{N}(y_i - \bar{Y})^2$$

Let us look closely at XY_{SP}, the numerator in our equation. First, notice that neither $(x_i - \bar{X})$ nor $(y_i - \bar{Y})$ is squared. So both could be positive, both could be negative, or one could be positive and one could be negative. This means the product of these can be either negative or positive. Now let us think about a few scenarios. Suppose there is a positive relationship between X and Y, so in general as X increases Y increases. That means in general we would expect that while we are looking at Xs below the mean of all Xs, we also expect the Ys we look at to be below the mean of all Ys. So we would get a negative times a negative resulting in a positive number. We would also expect that when we examine Xs above the mean of all Xs, we would see Ys that are also above the mean of all Ys. So we would get a positive times a positive, resulting in yet another positive number. In the end we should get a sum of mostly positive numbers. If you go through the same steps for a hypothetical negative relationship you would end up summing mostly negative numbers, resulting in a relatively large negative number. The numerator gives us our positive or negative association. If there is no relationship between X and Y, we expect get some negative products and some positive products, the sum of which will cancel many of these values out.

In the denominator, $(x_i - \overline{X})$ and $(y_i - \overline{Y})$ are both squared before being multiplied together—resulting in a positive number. This means the denominator is always positive, and $(x_i - \overline{X})$ and $(y_i - \overline{Y})$ cannot cancel each other out when these products are summed.

If X and Y have a strong relationship, the absolute value of the observed covariance will be close to the maximum possible positive covariance, yielding a r close to 1 or -1.

Compute correlation coefficient:

	x_i	y_i	$\overline{X}$	$\overline{Y}$	$(x_i - \overline{X})$	$(y_i - \overline{Y})$	$(x_i - \overline{X})(y_i - \overline{Y})$	$(x_i - \overline{X})^2$	$(y_i - \overline{Y})^2$
	1350	3.6	1353	3.5	-3	.1	-.3	9	.01
	1510	3.8	1353	3.5	157	.3	47.1	24649	0.09
	1420	3.7	1353	3.5	67	.2	13.4	4489	0.04
	1210	3.3	1353	3.5	-143	-.2	28.6	20449	0.04
	1250	3.9	1353	3.5	-103	.4	-41.2	10609	.16
	1300	3.4	1353	3.5	-53	-.1	5.3	2809	0.01
	1580	3.8	1353	3.5	227	.3	68.1	51529	.09
	1310	3.7	1353	3.5	-43	.2	-8.6	1849	0.04
	1290	3.5	1353	3.5	-63	0	0	3969	0
	1320	3.4	1353	3.5	-33	-.1	3.3	1089	0.01
	1490	3.8	1353	3.5	137	.3	41.1	18769	0.09
	1200	3.0	1353	3.5	-153	-.5	76.5	23409	.25
	1360	3.1	1353	3.5	7	-.4	-2.8	49	0.16
Totals	-	-	-	-	-	-	230.5	163677	0.99

$163677 \cdot .99 = 162040.23$, the square root of 162040.23 is 402.54, $230.5 / 402.54 = 0.57$, so $r = 0.57$.

We still don't know if r is significant. The null hypothesis is that there is no relationship between X and Y. We test this by using the shuffle method to break any relationship that may exist between the two, and see what the chances are that we get an r greater than or equal to .57.

48

4.8.3 Pseudocode & code

All code was tested using Python 2.3.
www.python.com

download code and input files here:
http://www.morganclaypool.com/userimages/ContentEditor/1216676640846/Statistics_is_Easy_Codes.zip

When we ran the CorrelationSig.py code on the example above we got:

```
Observed r: 0.58
151 out of 10000 experiments had a r greater than or equal to 0.58
Probability that chance alone gave us a r greater than or equal to 0.58 is
0.02
```

All code was tested using Python 2.3.
www.python.com
download code and input files here:
http://www.morganclaypool.com/userimages/ContentEditor/1216676640846/Statistics_is_Easy_Codes.zip

When we ran the CorrelationConf.py code on the example above we got:

```
Observed r: 0.58
We have  90.0 % confidence that the true r is between: 0.50 and 0.65
***** Bias Corrected Confidence Interval *****
We have 90.0 % confidence that the true r is between: 0.49 and 0.64
```

Correlation does not imply causation. Something other than X (but related to X) could be caus-ing the changes in Y. So we can use X to predict Y, but a change in X does not necessarily cause a change in Y. A classic example is that, before the polio vaccine, there was a positive correlation between sales of soda and the outbreak of polio. This doesn't mean that soda caused polio. So what was going on? It turned out that polio spread more easily in the summer when many people played together at pools and on the beach. That was also the time of year when soda sales went up. It is important to be careful when making assumptions about what causes what.

4.9   Multiple Regression

See the **CART** algorithm.

4.10 Multiple Testing

4.10.1 Why and when

Any time you measure the behavior of multiple entities (e.g., genes, plants, or other experimental subjects) $e1, ...e_n$, in an experiment, there is a chance that you will get a behavior that differs substantially from what is expected, at least for some of the entities. For example, if you do 10,000 experiments, and the probability of getting the unusual behavior is even as small as 0.05%, then you would actually expect to get about 5 entities displaying the unusual behavior. When you do a large number of tests, you run the risk of deciding the unusual behavior you come across is significant, when in fact it is expected. So how do you deal with this problem? There are several ways to handle it, and which you choose depends partly on whether it is more important to for you to be precise (few false positives) or have high recall (few false negatives). Use the Bonferroni correction to avoid false negatives (see Sec. 5.10.2 below).

4.10.2 Family Wise Error Rate

The family wise error rate (FWER) is the probability of getting a **type I error** (deciding that the null hypothesis is false when it is true) at least once when doing multiple significance tests.

Remember that the p-value is defined as the probability that chance alone would yield a test statistic less likely than the observed given that the null hypothesis is true. So in other words, the p-value measures the chance that for an *individual* test, the null hypothesis would be rejected when it is in fact true.

Use the Bonferroni correction to get a family-wise error rate less than q, by rejecting a null hypothesis only if the p-value multiplied by the number of experiments (significance tests run), is less than q.

By doing this, we have make it much harder for any individual significance test to reach significance. Say we want our family wise error rate to be less than 0.05 and we are running 10,000 tests (a number of tests not uncommon in biology). If we have a p-value of 0.0002, $(0.0002 \cdot 10,000) = 2$, and 2 is not less than 0.05, so we cannot reject the null hypothesis even though we have a very small p-value. This is a much more stringent test for significance. By decreasing the likelihood of **type I error**s, we increased the likelihood of **type II error**s, so we make it more likely that for an individual test we do not reject the null hypothesis when in fact we should.

4.10.3 False Discovery Rate

The false discovery rate (FDR) controls the expected proportion of type I errors when doing multiple significance tests. Frequently it is preferable to have some false positives (type I errors) in the results than to have false negatives (type II errors). For example, say you have an experiment in which you want to find the genes most likely to be active in breast cancer. So you are looking at tens of thousands of genes in maybe 100 patients. It is probably better to include a few genes that are not actually involved in breast cancer, than to exclude some that are. Using the false discovery rate you can say "here are the genes we predict are involved in breast cancer, with a FDR of 0.3." So if you have selected 10 genes, you expect 3 of them to be false positives.

The first step of the approach given by Benjamini and Hochberg is to order the p-values of all the experiments:

$$P_{(1)} \leq \ldots \leq P_{(m)}$$

where m is the total number of experiments. Notice that the first p-value is the smallest, and therefore the most significant.

The number of experiments where we reject the null hypothesis is given by:

$$t = \max \left\{ i : P(i) \leq q\left(\frac{i}{m}\right) \right\}$$

where q is the target false discovery proportion and m is the total number of significance tests. i is the number of experiments where we reject the null hypothesis. The formula can be read as: our cutoff t is the maximum i where $P(i)$ is less than or equal to (i / m) times q.

Case Study: New Mexico's 2004 Presidential Ballots

In our experience, knowing how to approach a problem statistically (understanding the data, figuring out what your question is, and deciding which test should be applied) is much more difficult for people than understanding how to run a pre-specified statistical test when given clean data. In practice, data is rarely as clean as data given in an example used for teaching purposes. Here we take you step by step through a complete data analysis on real data.

New Mexico was one of the key states in the 2004 presidential election and was an important victory for George W. Bush. Because 2004 was a close election and one of the first that made heavy use of electronic voting machines, many people wondered whether one voting machine was clearly better than another. Of course there are many criteria relevant to quality, one being mistakes in the recording of votes. That is the one we discuss here. We do not by contrast discuss the potential for manipulating a vote once inside the voting machine. We analyze the effects of voting machines and ethnicity in those elections.

5.1   Take a close look at the data

First we want to try to understand our data which can be downloaded from:

http://www.morganclaypool.com/userimages/ContentEditor/1214542013584/NMStateData.zip

We must clearly define everything and figure out not just what a term means, but exactly how it was defined during data collection.

Each ballot cast could have been done in one of three ways: early vote, election day vote, and absentee vote. We call these *voting types*.

For holding elections, each county in a state is divided into precincts. Each county can choose the voting technology it uses. Often, a county uses different voting technologies for each of the three voting types specified above.

For each voting type, there are several possible sources of error. An *undervote* is a ballot that does not contain a selection for president. The voter may omit the vote intentionally or the undervote may result from a failure of a voting machine.

A *phantom vote* is a vote for president where there was no actual ballot cast. It can only be a mistake; there should never be more votes for president than ballots cast.

The following is a contrived scenario:

```
100 Ballots
20  undervotes
      11 people opted not to vote for president,
      but did vote in the election
      9 votes for president were lost
5   phantom votes
```

This means that of the 89 people who chose to vote for a presidential candidate, 9 had their votes lost. In addition, 5 votes for president were recorded that shouldn't have been.

Since we don't have the original data, just the summary information that was recorded, the above scenario would show the following result:

```
100 Ballots
85  Votes Cast for President
  (80 successfully recorded real votes + 5 phantom votes = 85 recorded votes for
president)
```

From this we would compute:

```
15   undervotes (85 - 100 = -15, which is less than 0, and 100 - 85 = 15)
 0   phantom votes (85 - 100 = -15, which is not greater than 0, so 0)
```

As you can see, when **undervotes** and **phantom votes** are counted together they cancel each other out. Also, we won't know if an undervote was a mistake or was intentional. In order to reduce the canceling out effect, we want to count undervotes and phantom votes separately. That is, instead of taking summary information for the entire state (where either all the phantom votes will be canceled out by the undervotes or vice versa), we count undervotes and phantom votes at the precinct level so that hopefully less canceling out will occur. Even better, since we have summary information for the three different **types of votes** by precinct, we can use this set of summary information. Ideally, we could get down to the smallest set, the ballot. A ballot can only either have a vote for president, not have a vote for president, or the ballot could be fake, and therefore a phantom vote. If we had this level of information no canceling out would occur (because one ballot cast for president cannot be both an undervote and a phantom vote). As mentioned, we can't know if an individual undervote is a mistake or not. However, according to Liddle et al., undervote rates over 2% usually warrant investigation and so we can keep an eye out for this at the precinct level.

The **undervotes** for one **voting type** (early votes) in one precinct (precinct A, an abstract precinct) were counted by:

```
Early Vote undervote in Precinct A =
      max(0, Total Early ballots cast in Precinct A
        - Total Early Presidential Votes cast in Precinct A)

Early Vote phantom votes in Precinct A =
      max(0, Total Early Vote Presidential Votes Cast for Precinct A
          - Total Early Vote Ballots Cast for Precinct A)
```

The above holds for each **voting type** and each precinct.

```
Total undervotes for Precinct A = Early Vote undervote + Election Day undervote +
Absentee undervote
```

```
Total phantom votes = Early Vote phantom vote + Election Day phantom vote +
Absentee undervote
```

If you are still confused about how **undervotes** and **phantom votes** can cancel each other out make up a few scenarios yourself and start counting.

Figure 5.1:   Percent undervotes of total ballots cast by precinct.

The data is in the form of fraction of **undervotes** for each precinct. These numbers are plotted on a scatter plot above. 1429 precincts reported ballots cast. Notice that according to this graph there is one precinct where 100% of the total ballots were undervotes. This is very suspicious, so we go back to the data to learn more. We find that in precinct 999 of Dona Ana county (a precinct composed of the county's overseas absentee ballots) all 207 of the 207 ballots cast were undervotes. This looks like an outlier, but we do not need to think about whether or not to remove it because we will do a **rank transformation** on the data. Why we do this is explained in Sec. 5.1.2 below.

Figure 5.2:   Percent undervotes of total ballots cast by precinct (outlier removed).

Here we have temporarily removed the (possible) outlier in order to see the spread of fraction of **undervote**s for each precinct more clearly. Notice that the scale of the graph has changed. Glancing at this graph it appears that more than half of the precincts in New Mexico had over 2% undervotes. Remember undervote rates over 2% usually warrant investigation. In fact, 819 precincts had more than 2% undervotes, and 32 precincts had more than 10% undervotes.

Along the *x*-axis of this graph are fraction of undervotes for ballots cast. Displayed are the number of precincts for each range of fraction of undervotes. Here we can see for the majority of the precincts between 1% and 2% of the ballots cast were undervotes. We can see that there are many precincts with no undervotes which means a normal approximation does not work (because, for starters, there is a spike at zero.).

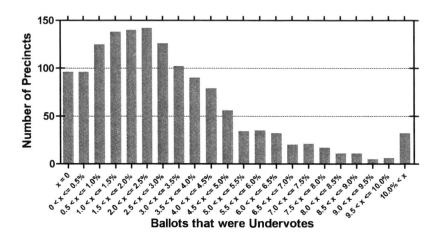

Figure 5.3: Ballots that were undervotes.

Next we look at the fraction of undervotes by voting type, machine type, and ethnic composition to see if anything interesting comes up.

> **BE CAREFUL**
>
> We shouldn't draw any conclusions without analyzing the data statistically.

In Fig. 5.4, it looks as if there is an association between undervotes and voting type (i.e., it looks as if election day voting is related to an increase in undervotes) as well as between undervotes and machine type in Fig. 5.5 (i.e., it looks as if use of the push button machines, Danaher Shouptronic and Sequoia Advantage, are related to an increase in undervotes).

- Push button DRE - Danaher Shouptronic and Sequoia Advantage
- Touch screen DRE - Sequoia Edge and ES&S iVotronic
- Optical scan (paper ballots) - Optech

Of course, the type of machine used to record votes is not the only thing that distinguishes New Mexico's precincts. Each precinct has a different composition of people as well. The precincts differ in ethnicity, education, income, distance of voter from polling station, etc. Any of these factors could be related to the difference in **undervote** rates between precincts.

Figure 5.6 shows New Mexico's ethnic composition. Anglos make up about 45%, Hispanics 42%, and Native Americans 9%. Blacks, Asians, and multi-racials, not shown, make up about 2%, 1%, and 1.5% of the population, respectively. The blue bar represents the overall undervote rate.

The next two graphs show the ethnic composition for selected precincts. The first shows the 10 precincts that had the smallest undervote rate (but did have at least one undervote each). The second shows the 10 precincts that had the largest undervote rate. In almost all of the precincts which

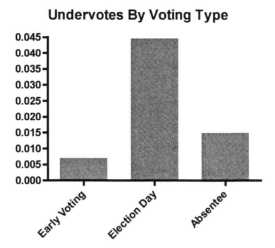

Figure 5.4: Undervotes by voting type.

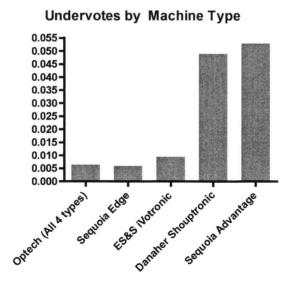

Figure 5.5: Undervotes by machine type.

had the smallest **undervote** rate, Anglos are by far the largest ethnic group and in almost all the precincts which had the largest undervote rate, Hispanics or Native Americans are the largest ethnic group. This is worth noting because in a democracy it is important not to disenfranchise part of the population.

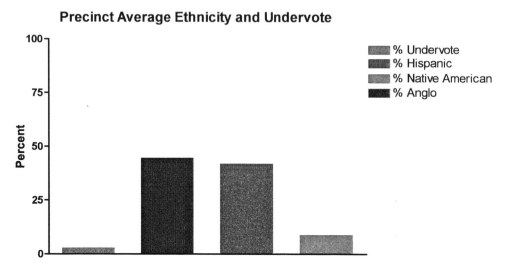

Figure 5.6:   New Mexico's ethnic composition.

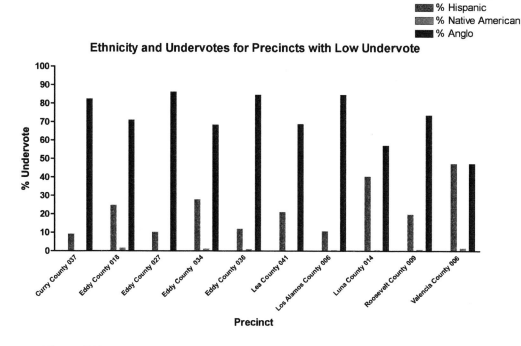

Figure 5.7:   Ethnicity and undervotes for precincts with low undervote.

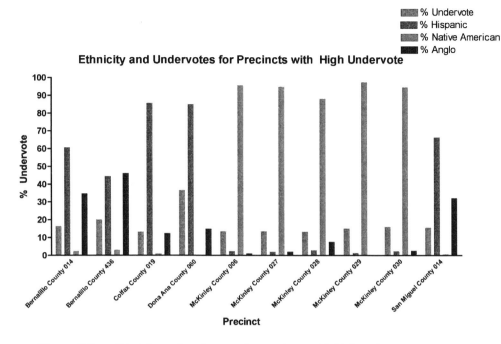

Figure 5.8:   Ethnicity and undervotes for precincts with high undervote.

See: http://www.democracyfornewmexico.com/democracy_for_new_mexico/files/ NewMexico2004ElectionDataReport-v2.pdf for a closer examination of this data.

5.1.1   What questions do we want to ask?

Of the many questions we could ask we are going to try to answer:

Is there a strong association between **undervotes** and some kind of machine?

5.1.2   How do we attempt to answer this question?

To decide whether or not there is a strong association between **undervotes** and some kind of machine, we will do a **one-way ANOVA**. We categorize the data on machine type. Therefore the groups are Push button DRE, Touch screen DRE, and Optical scan. The independent variable is the undervote rank.

We only look at election day data to eliminate **voting type** as a factor influencing **undervote** rate. Any precinct with 0 election day ballots is excluded from this test. We look at the fraction of undervotes in each precinct rather than the number of undervotes in each precinct, because 10 undervotes in a precinct where 20 ballots were submitted is very different from 10 undervotes out of 1,000 ballots. We do a rank transformation on the undervote rate to avoid having any outliers significantly distort the result. To assign rankings, the largest undervote rate was ranked 1, the second largest 2, etc. and the undervote rates for each precinct were sorted in descending order (largest first, smallest last). If multiple precincts had the same undervote rate, these precincts were all assigned the same rank. If there are x precincts with undervote rate u and r is the next unused rank, all x precincts are assigned rank $(r + ((r - 1) + x)) / 2$. The next unused rank is then $r + x$. For example, if ranks 1 through 3 have already been assigned, and the next 3 precincts have the same undervote rate, all 3 precincts are ranked $(4 + (3 + 3)) / 2 = 5$, and the next precinct is ranked 7 (assuming it has a unique undervote rate).

Pseudocode & code

All code was tested using Python 2.3.
www.python.com

download code and input files here:
http://www.morganclaypool.com/userimages/ContentEditor/1216676640846/Statistics_is_Easy_Codes.zip

When we ran the OneWayAnovaSig.py code on the example above we got:

```
Observed F-statistic: 217.29
0 out of 10000 experiments had a F-statistic greater than or equal to 217.29
Probability that chance alone gave us a F-statistic of 217.29 or more is 0.0
```

OK, the difference may be significant, but is it large? Does it matter?

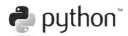

All code was tested using Python 2.3.
www.python.com
download code and input files here:
http://www.morganclaypool.com/userimages/ContentEditor/1216676640846/Statistics_is_Easy_Codes.zip

When we ran the OneWayAnovaConf.py code we got:

```
Observed F-statistic: 217.29
We have 90.0 % confidence that the true F-statistic is between: 179.85 and 260.16
***** Bias Corrected Confidence Interval *****
We have 90.0 % confidence that the true F-statistic is between: 178.74 and 258.85
```
We conclude that machine type was a significant *predictor* of **undervotes**. However, this does not necessarily mean that the machine type *caused* the undervote. The undervote could be caused by another factor that happens to be associated with machine type. For example, all precincts in a county use the same machine type on election day and it could be that county demographics are the cause of the difference in undervote numbers.

5.1.3   Next: effect of ethnicity for each machine type

We concluded that machine type is a significant predictor of **undervotes**, but since machine type is uniform across a county, we next try to remove the effect of county demographics. There are lots of variables related to county that we could examine, but since we have ethnicity information from the 2000 Census that is what we will look at.

We will do three **linear regressions**, one for each machine type. Each regression will compare ethnicity, in this case we measure the percent of the county that is minority (i.e., non-Anglo), to undervote rate.

First we look at push-button machines.

Pseudocode & code

All code was tested using Python 2.3.
www.python.com

download code and input files here:
http://www.morganclaypool.com/userimages/ContentEditor/1216676640846/Statistics_is_Easy_Codes.zip

When we ran the RegressionSig.py code we got:

```
Line of best fit for observed data:  y' = 0.0681 x +  1.5328
0 out of 10000 experiments had a slope greater than or equal to 0.0681 .
The chance of getting a slope greater than or equal to 0.0681 is 0.0 .
```

The slope is very small so a change in ethnicity is only slightly related to a change in undervote rate. However, the result is significant.

Next we get a confidence interval for the slope.

Pseudocode & code

All code was tested using Python 2.3.
www.python.com

download code and input files here:
http://www.morganclaypool.com/userimages/ContentEditor/1216676640846/Statistics_is_Easy_Codes.zip

When we ran the RegressionSig.py code we got:

```
Line of best fit for observed data:  y' = 0.0681 x +  1.5328
We have 95.0 % confidence that the true slope is between: 0.0614 and 0.0749
***** Bias Corrected Confidence Interval *****
We have 95.0 % confidence that the true slope is between: 0.0646 and 0.0715
```

Notice that we have 95% confidence that the slope is positive. Next we check how well ethnicity predicts undervote rate.

Pseudocode & code

All code was tested using Python 2.3.
www.python.com

download code and input files here:
http://www.morganclaypool.com/userimages/ContentEditor/1216676640846/Statistics_is_Easy_Codes.zip

When we ran the CorrelationSig.py code we got:

```
Observed r: 0.52
0 out of 10000 experiments had a r greater than or equal to 0.52
Probability that chance alone gave us a r greater than or equal to 0.52 is 0.00
```

Ethnicity isn't a great predictor of undervote rank, but it isn't bad.
Next we do the same for optical machines.

Pseudocode & code

All code was tested using Python 2.3.
www.python.com

download code and input files here:
http://www.morganclaypool.com/userimages/ContentEditor/1216676640846/Statistics_is_Easy_Codes.zip

When we ran the RegressionSig.py code we got:

```
Line of best fit for observed data:  y' = 0.0249 x +  0.5090
4 out of 10000 experiments had a slope greater than or equal to 0.0249 .
The chance of getting a slope greater than or equal to 0.0249 is 0.0004 .
```

The slope is very small so a change in ethnicity is only slightly related to a change in undervote rate. However, the result is significant.
Next we get a confidence interval for the slope.

Pseudocode & code

All code was tested using Python 2.3.
www.python.com

download code and input files here:
http://www.morganclaypool.com/userimages/ContentEditor/1216676640846/Statistics_is_Easy_Codes.zip

When we ran the RegressionConf.py code we got:

```
Line of best fit for observed data:  y' = 0.0249 x +  0.5090
We have 90.0 % confidence that the true slope is between: 0.0137 and 0.0367
***** Bias Corrected Confidence Interval *****
We have 90.0 % confidence that the true slope is between: 0.0138 and 0.0368
```

Again, notice that we have 95% confidence that the slope is positive. Next we check how well ethnicity predicts undervote rate.

Pseudocode & code

All code was tested using Python 2.3.
www.python.com

download code and input files here:
http://www.morganclaypool.com/userimages/ContentEditor/1216676640846/Statistics_is_Easy_Codes.zip

When we ran the CorrelationSig.py code we got:

```
Observed r: 0.24
2 out of 10000 experiments had a r greater than or equal to 0.24
Probability that chance alone gave us a r greater than or equal to 0.24 is 0.00
```

Ethnicity isn't a great predictor of undervote rank.
Finally, we do the same for touch screen machines.

Pseudocode & code

All code was tested using Python 2.3.
www.python.com

download code and input files here:
http://www.morganclaypool.com/userimages/ContentEditor/1216676640846/Statistics_is_Easy_Codes.zip

When we ran the RegressionSig.py code we got:

```
Line of best fit for observed data:  y' = -0.0659 x +  5.5577
0 out of 10000 experiments had a slope less than or equal to -0.0659 .
The chance of getting a slope less than or equal to -0.0659 is 0.0 .
```

The slope is very small so a change in ethnicity is only slightly related to a change in undervote rate. However the result is significant.

Next we get a confidence interval for the slope.

Pseudocode & code

All code was tested using Python 2.3.
www.python.com

download code and input files here:
http://www.morganclaypool.com/userimages/ContentEditor/1216676640846/Statistics_is_Easy_Codes.zip

When we ran the RegressionConf.py code we got:

```
Line of best fit for observed data:  y' = -0.0659 x +  5.5577
We have 95.0 % confidence that the true slope is between: -0.0927 and -0.0421
***** Bias Corrected Confidence Interval *****
We have 95.0 % confidence that the true slope is between: -0.0790 and -0.0540
```

Notice that here we have 95% confidence that the slope is *negative*. Next we check how well ethnicity predicts undervote rate.

Pseudocode & code

All code was tested using Python 2.3.
www.python.com

download code and input files here:
http://www.morganclaypool.com/userimages/ContentEditor/1216676640846/Statistics_is_Easy_Codes.zip

When we ran the CorrelationSig.py code we got:

```
Observed r: -0.52
0 out of 10000 experiments had a r less than or equal to -0.52
Probability that chance alone gave us a r less than or equal to -0.52 is 0.00
```

Ethnicity isn't a great predictor of undervote rank, but it isn't bad.

Now let's look at the results of this analysis in a graph in Figure 5.9.

If the slopes of these lines had all been uniformly negative or positive, we would have suspected that ethnicity had a strong effect on undervote counts. This does not hold. So, ethnicity is no proxy for tendency to undervote. The next question is whether the voting machine itself could determine the tendency to undervote. To minimize the effect of ethnicity, we can split the data into quadrants based on percent minority, and then for each quadrant ask if the machine types are significantly different. The purpose of doing this would be to look at sections of the data in which percent minorty does not vary much. At the end of this exercise we may be able to say, in the case where percent minority is between *a* and *b*, push button, touch screen, and optical scan machines do/do not significanlty differ.

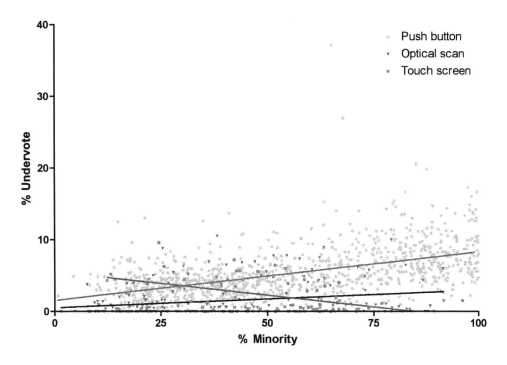

Figure 5.9: Percent minority comparison between push button, touch screen, and optical scan machines.

Since we want to determine if the three machine types significantly differ in each of these four cases, we will do four **one-way ANOVAs**. First we look at precincts that have a percent minority greater than or equal to zero and less than 25.

Pseudocode & code

All code was tested using Python 2.3.
www.python.com

download code and input files here:
http://www.morganclaypool.com/userimages/ContentEditor/1216676640846/Statistics_is_Easy_Codes.zip

When we ran the OneWayAnovaSig.py code we got:

```
Observed F-statistic: 61.12
0 out of 10000 experiments had a F-statistic greater than or equal to 61.12
Probability that chance alone gave us a F-statistic of 61.12 or more is 0.0
```

OK, the difference may be significant, but is it large? Does it matter?

Pseudocode & code

All code was tested using Python 2.3.
www.python.com

download code and input files here:
http://www.morganclaypool.com/userimages/ContentEditor/1216676640846/Statistics_is_Easy_Codes.zip

When we ran the OneWayAnovaConf.py code we got:

```
Observed F-statistic: 61.12
We have 90.0 % confidence that the true F-statistic is between: 47.68 and 79.64
***** Bias Corrected Confidence Interval *****
We have 90.0 % confidence that the true F-statistic is between: 45.33 and 76.74
```

Now we look at the precincts that have percent minority greater than or equal to 25 and less than 50.

Pseudocode & code

All code was tested using Python 2.3.
www.python.com

download code and input files here:
http://www.morganclaypool.com/userimages/ContentEditor/1216676640846/Statistics_is_Easy_Codes.zip

When we ran the OneWayAnovaSig.py code we got:

```
Observed F-statistic: 36.21
0 out of 10000 experiments had a F-statistic greater than or equal to 36.21
Probability that chance alone gave us a F-statistic of 36.21 or more is 0.0
```

OK, the difference may be significant, but is it large? Does it matter?

Pseudocode & code

All code was tested using Python 2.3.
www.python.com

download code and input files here:
http://www.morganclaypool.com/userimages/ContentEditor/1216676640846/Statistics_is_Easy_Codes.zip

When we ran the OneWayAnovaConf.py code we got:

```
Observed F-statistic: 36.21
We have 90.0 % confidence that the true F-statistic is between: 21.70 and 57.96
***** Bias Corrected Confidence Interval *****
We have 90.0 % confidence that the true F-statistic is between: 19.83 and 55.18
```

Next we look at precincts where the percent minority is greater than or equal to 50 and less than 75.

Pseudocode & code

All code was tested using Python 2.3.
www.python.com

download code and input files here:
http://www.morganclaypool.com/userimages/ContentEditor/1216676640846/Statistics_is_Easy_Codes.zip

When we ran the OneWayAnovaSig.py code we got:

```
Observed F-statistic: 61.70
0 out of 10000 experiments had a F-statistic greater than or equal to 61.70
Probability that chance alone gave us a F-statistic of 61.70 or more is 0.0
```

OK, the difference may be significant, but is it large? Does it matter?

Pseudocode & code

All code was tested using Python 2.3.
www.python.com

download code and input files here:
http://www.morganclaypool.com/userimages/ContentEditor/1216676640846/Statistics_is_Easy_Codes.zip

When we ran the OneWayAnovaConf.py code we got:

```
Observed F-statistic: 61.70
We have 90.0 % confidence that the true F-statistic is between: 43.77 and 87.03
***** Bias Corrected Confidence Interval *****
We have 90.0 % confidence that the true F-statistic is between: 42.12 and 83.99
```

Finally, we look at precincts where the percent minority is greater than or equal to 75 and less than 100.

Pseudocode & code

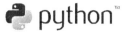

All code was tested using Python 2.3.
www.python.com

download code and input files here:
http://www.morganclaypool.com/userimages/ContentEditor/1216676640846/Statistics_is_Easy_Codes.zip

When we ran the OneWayAnovaSig.py code we got:

```
Observed F-statistic: 74.62
0 out of 10000 experiments had a F-statistic greater than or equal to 74.62
Probability that chance alone gave us a F-statistic of 74.62 or more is 0.0
```
   OK, the difference may be significant, but is it large? Does it matter?

Pseudocode & code

All code was tested using Python 2.3.
www.python.com

download code and input files here:
http://www.morganclaypool.com/userimages/ContentEditor/1216676640846/Statistics_is_Easy_Codes.zip

When we ran the OneWayAnovaConf.py code we got:

```
Observed F-statistic: 74.62
We have 90.0 % confidence that the true F-statistic is between: 45.87 and 116.16
***** Bias Corrected Confidence Interval *****
We have 90.0 % confidence that the true F-statistic is between: 43.64 and 112.41
```

   We conclude that in all four quadrants the machine types are significantly different. Note that this does not mean that in any given quadrant machine type a is significantly different from machine type b, only that the three machines types differ. To determine if any two machine types are significantly different in a quadrant, we must compare those two machine types. In addition, we have not said which machine type performs the best in each quadrant. To do that we would have to compare the means.

   Remember we have not shown that a change in percent minority *causes* the change in the expected undervote rate for a particular machine, we are claiming only that a change in percent minority is correlated with a change in the undervote rate for a particular machine. It is an important distinction to make. Any number of things could be causing the change in undervote rate for a machine type: income, distance from polling station, quality of public education, comfort level with computers, etc.

5.1.4 Techniques we have used

To measure the difference in undervotes for the three machine types we used a **One-way ANOVA**. We did a rank transformation on the undervote rate to avoid having outliers distort the result of the one-way ANOVA.

Since each county in New Mexico used one machine type on election day, instead of the machine types being randomly distributed across the state, we attempted to remove the effect of a county demographic, ethnicity, to compare machine type and undervote rate. To get an idea of how ethnicity is related to the relationship between undervotes and machine types, we ran three **linear regressions**, one for each machine type. We then overlaid the three regression lines in a graph to compare them.

The effects of machine type and ethnicity are interrelated. To see if the three machine types significantly differed in subsections of the population where the variance in percent minority is smaller, we split the precincts up into four quadrants based on percent minority and then ran four **One-way ANOVAs**.

5.1.5 What did we find out?

One-way ANOVA suggests that certain machine types lead to more undervotes, but which machine type performs the best depends on the demographics of the area. In fact, the voters using touch screen machines suffered the least undervotes when the percent minority (non-Anglo) was low, but, in all other cases, voters using push-button machines suffered the least undervotes. So, it is not true that "a single machine type consistently more likely to yield undervotes over all ethnicity levels."

There would still be other questions to ask such as the degree of repair of the voting machines in the different countries, the model, the number of machines, the length of waiting lines. We do not have that information, but if you do, then you can use a similar analysis.

References

Andrews, D. F. F., Bickel, P. J., Hampel, F. R., Tukey, J. W., and Huber, P. J., *Robust Estimates of Location: Survey and Advances*, Princeton University Press, Princeton, 1972.

Bruce, P. C. and Simon, J. L., "The Statistical Results of Resampling Instruction: Evaluations of Teaching Introductory Statistics via Resampling," *Resampling Stats*. http://www.resample.com/content/teaching/texts/i-1reslt.txt

Campbell, M. K. and Torgerson, D. J., "Bootstrapping: estimating confidence intervals for cost-effectiveness ratios," *Quarterly Journal of Medicine,* 92:177-182, 1999. http://qjmed.oxfordjournals.org/cgi/content/full/92/3/177

Chiaromonte, F., "Statistical Analysis of Genomics Data," New York University Courant Institute of Mathematical Sciences, New York, http://www.cims.nyu.edu/~chiaro/SAGD_05

(An excellent introduction to resampling in the context of biological problems.)

Davison, A.C. and Hinkley, D. V., *Bootstrap Methods and their Application*. Cambridge University Press, Cambridge, 1997.

"Distribution Tables," StatSoft. http://www.statsoft.com/textbook/sttable.html

Easton, V. J. and McColl, J. H., "Design of Experiments and ANOVA," *Statistics Glossary*. http://www.cas.lancs.ac.uk/glossary_v1.1/dexanova.html

Good, P. I., *Resampling Methods: A Practical Guide to Data Analysis*, Birkhäuser, Boston, 2006.

Howell, D., "Resampling Statistics: Randomization and the Bootstrap." The University of Vermont, Burlington, VT. http://www.uvm.edu/~dhowell/StatPages/Resampling/Resampling.html

Liddle, E., Mitteldorf, J., Sheehan, R. G., and Baiman, R., "Analysis of Undervotes in New Mexico's 2004 Presidential Ballots," National Election Data Archive. Jan. 2005. http://uscountvotes.org/ucvAnalysis/NM/NMAnalysis_EL_JM.pdf

Lowry, R., *Concepts and Applications of Inferential Statistics*. http://faculty.vassar.edu/lowry/webtext.html

Lunneborg, C. E., *Data Analysis by Resampling: Concepts and Applications*, Duxbury, 2000.

Lunneborg, C. E., "Modeling Experimental and Observational Data", iUniverse.com, 2000.

"Muriel Bristol." Wikipedia. http://en.wikipedia.org/wiki/Muriel_Bristol

"Outlier." Wikipedia. http://en.wikipedia.org/wiki/Outlier

Sen, S., "Multiple Comparisons and the False Discovery Rate." University of California San Francisco. http://www.biostat.ucsf.edu/biostat/sen/papers/fdr.pdf

Taleb, N., "Learning to Expect the Unexpected." *Edge*. 14 Apr. 2004. http://www.edge.org/3rd_culture/taleb04/taleb_index.html

Theisen, E. and Stewart, W., "Summary Report on New Mexico State Election Data." Democracy for New Mexico. www.HelpAmericaRecount.Org. January 4, 2005. http://www.democracy-fornewmexico.com/democracy_for_new_mexico/files/NewMexico2004ElectionDataReport-v2.pdf

Walters, S. J., "Sample size and power estimation for studies with health related quality of life outcomes: a comparison of four methods using the SF-36," *Health Qual Life Outcomes*, 2:26, 2004. http://www.pubmedcentral.nih.gov/articlerender.fcgi?artid=421748

Weaver, B., Statistics Notes. http://www.angelfire.com/wv/bwhomedir/notes.html

Bias Corrected Confidence Intervals

Our method of computing **confidence intervals** works well when our original sample is the median of the bootstrap values. When this is not the case, we say the data is "biased." Bias often arises because extreme values in the data are on one side only, e.g., a collection of wind speed values including a few from hurricanes. We can correct for bias by computing how far the original sample value is from the mean of our bootstrap values, and adjusting our interval accordingly. Resampling pioneer Bradley Efron calls this a "bias corrected confidence interval"(see http://qjmed.oxfordjournals.org/cgi/content/full/92/3/177).

First, we find the fraction of **bootstraps** below the value computed from the original sample (that value is hereafter called the *original sample estimate*). To do this, order all the bootstrap sample estimates, count the number of them that are below the original sample estimate, then divide that by the total number of bootstrap estimates. So, if we did 10,000 bootstraps and 4,500 of them have a sample estimate below our original sample estimate, then 4,500/10,000 = 0.45 (45% of the bootstrap estimates are below our whole estimate). Intuitively, this means that the estimate based on the original sample is less than the median of the bootstrapping. So, we'll have to adjust the confidence interval downwards because the estimate based on the original sample should get more weight than the bootstraps. Intuition has told us in which direction to adjust the confidence interval. A slightly involved statistical algorithm will now tell us how much. The python code associated with each statistic discussed later implements this algorithm as one option.

Imagine a normal curve. The mean is the same as the median that is greater than 50% of the values. A point that is greater than only 45% of the values is shown just to the left. A fraction 0.05 of all values are in between (see Fig. A.1). We can find out how many standard deviations away the 0.45 point is by consulting the ϕ table below.

The left column of the table represents fractions of standard deviations in increments of 0.1. The column headers represent additional fractions of standard deviations in increments of 0.01. Thus, the cell where the row 0.1 and the column 0.02 intersect corresponds to 0.12 standard deviations. In that cell, we find 0.0478. This expresses the function $\phi (0.12) = 0.0478$. So, under a normal assumption, a value that is greater than $0.5 + 0.0478$ ($= 0.5478$) of the values is 0.12 standard deviations above the mean. A value that is greater than $0.5 + 0.0517$ ($= 0.5517$) is 0.13 standard deviations above the mean. Interpolating, we see that a value that is greater than 0.55 of the values is approximately 0.125 standard deviations above the mean. In our case, we are talking about a value that is greater than 0.45 of the values (that is, a value 0.05 *below* the mean) so its value in standard deviations is symmetrically negative, i.e., -0.125. We call that value the bias correcting constant z_0. In summary, the table computes the function ϕ from standard deviations (in bold) to fractions of values (in plaintext in the cells) above the median. It is symmetric for values below the median.

If you now want a 95% confidence interval, the fraction of bootstraps that we are omitting is 0.05, so the fraction of bootstraps that we are omitting on each end is half that or .025, and we want to go from 2.5% to 97.5%. So, 47.5% of the values should be directly below the mean, and 47.5%

Figure A.1: Standard deviations from mean.

Area between the mean and a value x standard deviations above the mean[*]

	0.00	0.01	0.02	0.03	0.04	0.05	0.06	0.07	0.08	0.09
0.0	0.0000	0.0040	0.0080	0.0120	0.0160	0.0199	0.0239	0.0279	0.0319	0.0359
0.1	0.0398	0.0438	**0.0478**	**0.0517**	0.0557	0.0596	0.0636	0.0675	0.0714	0.0753
0.2	0.0793	0.0832	0.0871	0.0910	0.0948	0.0987	0.1026	0.1064	0.1103	0.1141
0.3	0.1179	0.1217	0.1255	0.1293	0.1331	0.1368	0.1406	0.1443	0.1480	0.1517
0.4	0.1554	0.1591	0.1628	0.1664	0.1700	0.1736	0.1772	0.1808	0.1844	0.1879
0.5	0.1915	0.1950	0.1985	0.2019	0.2054	0.2088	0.2123	0.2157	0.2190	0.2224
0.6	0.2257	0.2291	0.2324	0.2357	0.2389	0.2422	0.2454	0.2486	0.2517	0.2549
0.7	0.2580	0.2611	0.2642	0.2673	0.2704	0.2734	0.2764	0.2794	0.2823	0.2852
0.8	0.2881	0.2910	0.2939	0.2967	0.2995	0.3023	0.3051	0.3078	0.3106	0.3133
0.9	0.3159	0.3186	0.3212	0.3238	0.3264	0.3289	0.3315	0.3340	0.3365	0.3389
1.0	0.3413	0.3438	0.3461	0.3485	0.3508	0.3531	0.3554	0.3577	0.3599	0.3621
1.1	0.3643	0.3665	0.3686	0.3708	0.3729	0.3749	0.3770	0.3790	0.3810	0.3830
1.2	0.3849	0.3869	0.3888	0.3907	0.3925	0.3944	0.3962	0.3980	0.3997	0.4015
1.3	0.4032	0.4049	0.4066	0.4082	0.4099	0.4115	0.4131	0.4147	0.4162	0.4177
1.4	0.4192	0.4207	0.4222	0.4236	0.4251	0.4265	0.4279	0.4292	0.4306	0.4319
1.5	0.4332	0.4345	0.4357	0.4370	0.4382	0.4394	0.4406	0.4418	0.4429	0.4441
1.6	0.4452	0.4463	0.4474	0.4484	0.4495	0.4505	0.4515	0.4525	0.4535	0.4545
1.7	0.4554	**0.4564**	0.4573	0.4582	0.4591	0.4599	0.4608	0.4616	0.4625	0.4633
1.8	0.4641	0.4649	0.4656	0.4664	0.4671	0.4678	0.4686	0.4693	0.4699	0.4706
1.9	0.4713	0.4719	0.4726	0.4732	0.4738	0.4744	**0.4750**	0.4756	0.4761	0.4767
2.0	0.4772	0.4778	0.4783	0.4788	0.4793	0.4798	0.4803	0.4808	0.4812	0.4817

2.1	0.4821	0.4826	0.4830	0.4834	0.4838	0.4842	0.4846	0.4850	0.4854	0.4857
2.2	0.4861	**0.4864**	0.4868	0.4871	0.4875	0.4878	0.4881	0.4884	0.4887	0.4890
2.3	0.4893	0.4896	0.4898	0.4901	0.4904	0.4906	0.4909	0.4911	0.4913	0.4916
2.4	0.4918	0.4920	0.4922	0.4925	0.4927	0.4929	0.4931	0.4932	0.4934	0.4936
2.5	0.4938	0.4940	0.4941	0.4943	0.4945	0.4946	0.4948	0.4949	0.4951	0.4952
2.6	0.4953	0.4955	0.4956	0.4957	0.4959	0.4960	0.4961	0.4962	0.4963	0.4964
2.7	0.4965	0.4966	0.4967	0.4968	0.4969	0.4970	0.4971	0.4972	0.4973	0.4974
2.8	0.4974	0.4975	0.4976	0.4977	0.4977	0.4978	0.4979	0.4979	0.4980	0.4981
2.9	0.4981	0.4982	0.4982	0.4983	0.4984	0.4984	0.4985	0.4985	0.4986	0.4986
3.0	0.4987	0.4987	0.4987	0.4988	0.4988	0.4989	0.4989	0.4989	0.4990	0.4990

*The normal distribution table comes from: http://www.statsoft.com/textbook/stathome.html?sttable.html&1.

above (because 50% - 2.5% = 47.5%). By looking at the table we find that .475 of the values lie 1.96 standard deviations below the mean, and .475 of the values lie 1.96 standard deviations above the mean. $Z_{\alpha/2}= -1.96$; $Z_{1-\alpha/2}= 1.96$. This is the standard normal deviate associated with $\alpha/2$.

The lower limit of the bias-corrected confidence interval is the value of the bootstrap estimate at the $0.5 + \phi(Z_{\alpha/2} + 2z_0)^{th}$ percentile and the upper limit is the $0.5 + \phi(Z_{1-\alpha/2} + 2z_0)^{th}$ percentile. From our example above we have $Z_{\alpha/2}= -1.96$ and $Z_{1-\alpha/2} = 1.96$. Remember that z_0 is the standard normal deviate corresponding to the proportion of bootstrap values below our original sample value. So in our example, $z_0 = -0.125$, $\phi(-1.96 + 2(-0.125)) = \phi(-2.21) = -0.4864$, $\phi(1.96 + 2(-0.125)) = \phi(1.71) = 0.4564$, and so we look at the bootstrap estimates at the $(0.5 + -0.4864)^{th} = 0.0136^{th}$ percentile position, and the $(0.5 + 0.4564)^{th} = 0.9546^{th}$ percentile position. Since we took 10,000 bootstraps, we have a 95% confidence interval from the bootstrap value at position 136 (we got 136 from 0.0136 · 10,000) to the bootstrap value at position 9564 (we got 9564 from 0.9564 · 10,000).

All code was tested using Python 2.3.
www.python.com

download code and input files here:
http://www.morganclaypool.com/userimages/ContentEditor/1216676640846/Statistics_is_Easy_Codes.zip

Here is an example run of this code, using the Diff2MeanConfCorr.py from our previous placebo versus drug example (found on p. 6):

```
Observed difference between the means: 12.97
We have 90.0 % confidence that the true difference between the means is between:
7.60 and 18.12
```

The reason it should be $+2 z_0$ can be explained intuitively as follows. Imagine a town in which there are relatively few very rich people and the rest are of modest means. Suppose the statistic is the mean. In this town the mean will be greater than the median. In that case the proportion p that are below the sample estimate will be > 0.5. So z_0 will be positive reflecting the fact that the mean is so many standard deviations above the median. (The percentile method would give the same result if the median and mean were the same value so z_0 were 0.) We use this in the calculation to give a little extra weight to those few rich people, because the original sample had them.